Design for Life

Desi

Our daily lives, the spaces we shape, and the ways we communicate, as seen through the collections of Cooper-Hewitt, National Design Museum.

gn for Life

by Susan Yelavich.
Edited by Stephen Doyle.
Designed and produced
by Drenttel Doyle Partners.

Cooper-Hewitt,
National Design Museum
and Rizzoli, New York

First published in the
United States of America in 1997 by
Cooper-Hewitt
National Design Museum
Smithsonian Institution
2 East 91st Street
New York, NY 10128
and
Rizzoli International Publications, Inc.
300 Park Avenue South
New York, NY 10010

Trade Edition ISBN 0-8478-2030-0
Museum Edition ISBN 0-910503-64-8 (pbk)
Museum Edition ISBN 0-910503-63-X (hc)

Library of Congress
Cataloguing-in-Publication Data
Cooper-Hewitt Museum
Design for life: our daily lives, the spaces we shape, and the ways we communicate, as seen through the collections of Cooper-Hewitt, National Design Museum / by Susan Yelavich; edited by Stephen Doyle; designed and produced by Drenttel Doyle Partners.
192 p. 22.2 x 29.3 cm.
ISBN 0-910503-63-X—
ISBN 0-910503-64-8 (pbk.)—
ISBN 0-8478-2030-0 (Rizzoli:pbk.)
1. Design—Catalogs. 2. Design, Industrial—Catalogs. 3. Design—New York (State)—New York—Catalogs. 4. Cooper-Hewitt Museum—Catalogs. I. Yelavich, Susan. II. Title.
NK460.N4C666 1997
745.2 0747471—dc21 96-53981
CIP

Color separations by Graphic Service, Milan
Grafiche Editoriali Padane, Cremona
Printed and bound in Italy

FRONT COVER

Atomic Clock
George Nelson Associates, USA, 1949.

Model #302 Telephone
Henry Dreyfuss, USA, 1937.

Design for "Sportshack"
Donald Deskey, USA, 1940.

BACK COVER

"El Dorado" Wallpaper
Eugène Ehrmann, George Zipélius, and Joseph Fuchs, France, 1848.

Radius Toothbrush
James O'Halloran, USA, 1983.

Man's Hat
France or Italy, 18th century.

INSIDE FRONT COVER

"Buttons" wallpaper
USA, 1952.

CONTENTS PAGE (opposite)

Toy Grasshopper
Europe, c. 1900.

INSIDE BACK COVER

"Woodpidgeon" wallpaper
Edward Bawden, England, 1927.

This publication celebrates the one hundredth anniversary of Cooper-Hewitt, National Design Museum.

FOREWORD: *Design for Life* draws on the experience of a century of collecting, documenting, and studying design. What follows is a step toward furthering that academic category, design history. The meaning of the word "design" has changed over the centuries. But design has existed since human beings picked up rocks and carved them for use as tools.

"Design" is both a verb and a noun. For this reason, the Museum has always been as interested in the process of design as in the end result. Those results—the ordinary objects that we all use to both survive and enjoy life—are preserved to understand the past and hopefully to make the future better.

The impulse to design varies according to needs and wishes, time and place, cultural and social conventions, materials and technology. As an active process involving inquiry, ingenuity, and inspiration, design can change and dramatically enhance the quality of life.

To help the public understand how design impacts on our lives every second of the day, this book is arranged thematically, inviting the reader to consider objects in the collection in the context of daily life. The rituals of our days and the ways in which we live our lives within society are the framework through which the Museum's broad and multifaceted collection is presented.

The National Design Museum is one of the largest repositories of design in the world, with a collection of nearly a quarter-million objects. From typewriters to teapots, architectural renderings to lace, wallpaper sample books to posters, the Museum's holdings are as varied as they are unusual. In collecting the products of humankind's inventive and creative energy, the Museum values "masterpiece" and "mass-produced" equally for their success in creative problem solving.

It is at once exhilarating and frustrating that *Design for Life* can illustrate only about one percent of the National Design Museum's collection. Thousands of other works might have been substituted for the examples shown here. What these pages reveal is only the tip of the iceberg, a sampling of the depth and variety of its holdings.

The core of the Museum collection was formed in the 19th and early 20th centuries by the granddaughters of industrialist Peter Cooper. Inspired by the Musée des Arts Décoratifs in Paris and the South Kensington Museum (now the Victoria and Albert) in London, the three Hewitt sisters—Sarah, Eleanor, and Amy—worked to create a "visual library" of design, establishing the collections as a "working" museum. Amassing a variety of examples of design that ranged from pieces of furniture to drawings, prints, and textiles, they established the Cooper Union Museum for the Arts of Decoration in 1897 for the purpose of raising standards of design in America.

In establishing perhaps the first design museum in the United States, the Hewitt sisters were uniquely interested in giving students, designers, and the general public ready access to the collections. The idea of the museum as a working laboratory has remained a distinguishing characteristic of the Museum's mission and is highlighted by the opening of the Design Resource Center in the fall of 1997.

The Hewitt sisters' vision guided the Museum's collecting interests throughout the first half of the 20th century. The resulting collection was augmented by gifts, bequests, and purchases over the years. Succeeding generations of curators have enlarged the scope of the collections with objects of contemporary design, all the while continuing to increase its impressive historical range.

In 1967, the collections of the Museum were transferred to the Smithsonian Institution and found a new home in the 1902 mansion that had belonged to Andrew Carnegie in New York City. The doors of the newly named Cooper-Hewitt, National Design Museum opened to the public in 1976, launching an enlightening and entertaining program that focused on the products and processes of design.

One of the most exciting aspects of our collection is that the individual objects continue to live. The Museum's buttons, jewelry, fans, books, posters, and trade catalogues are a continuing source of inspiration for scholars, designers, students, and the public.

The collection is divided into four discrete curatorial areas: applied arts and industrial design, textiles, wallcoverings, and drawings and prints, each overseen by a curatorial team responsible for its care, documentation, and interpretation. Extensive curatorial research for this volume was provided by: Deborah Shinn, Assistant Curator (and Acting Head), Applied Arts and Industrial Design; Milton Sonday, Curator, Gillian Moss, Assistant Curator, and Lucy Commoner, Conservator, Textiles; Joanne Warner, Assistant Curator, Wallcoverings; Marilyn Symmes, Curator, Gail S. Davidson, Assistant Curator, and Elizabeth H. Marcus, Curatorial Assistant, Drawings and Prints. Their contributions provided the foundation for the book and are gratefully acknowledged, as are those of Russell Flinchum with regard to the Henry Dreyfuss material. Ellen Lupton, Curator, Contemporary Design; Egle Žygas, Program Coordinator, Education; Tracy Myers, Special Assistant to the Assistant Director for Public Programs; and Caroline

Mortimer, Special Assistant to the Director, served as readers for the manuscript, and their responses are deeply appreciated. Additional thanks go to the professional staff that supported the departmental work and administered the many details that a book as large as this entailed. Special mention must be made of Linda J. Herd's and Hilda Wojack's dedication to the enormous task of organizing the book's photography and captions. My thanks are also extended to the editor of this publication, Lorraine Karafel.

The Museum's Library is an unparalleled resource for information about design and is administered as part of the Smithsonian Institution Libraries. Chief Librarian Stephen Van Dyk has overseen the use of the Library materials in this publication, as well as the Museum's extensive archives, assisted by Janis Staggs-Flinchum.

We are especially grateful to the J.M. Kaplan Fund, without whose generous support this publication would not have been possible. Additional funding was provided by the Lisa Taylor Fund, National Design Museum.

Susan Yelavich, Assistant Director for Public Programs, deserves special mention for her critical, all-encompassing role in making the publication a reality. Susan has worked tirelessly with the curators and designers in this effort. A critical thinker, articulate and impassioned, Susan has given her all to every aspect of this project, and I am greatly indebted to her.

I am grateful to Drenttel Doyle Partners for their design of the book. This is the crowning achievement of the relationship between the design firm and the Museum. In 1994, Drenttel Doyle Partners created a new graphic identity for the National Design Museum. Since that time, they have been active participants in many publication and exhibition projects. This book is itself an artifact of design—the result of a unique collaboration between its designer, Stephen Doyle, and its author, Susan Yelavich. It has been a co-production from the outset, acknowledging that the best results come when design is integral to content—from inception to production.

As objects of design, our collections can only be fully understood in light of the context of their use. As the National Design Museum, we are charged with telling these stories in the hope that future generations will shape their world with care, protect our planet, make people's lives better, and design everything with imagination and beauty, as well as universal function.

Dianne H. Pilgrim
Director

The collections of Cooper-Hewitt, National Design Museum are the material evidence of our lives. They reveal the character of human nature from its most selfless to vainglorious. Each object tells a story about us and about the perpetual process of design that is central to our existence. How can we explain this restless urge to reinvent the world, generation after generation, from culture to culture? We survive if we have food, clothing, and shelter, but we create families, communities, and civilizations by imbuing these basic requirements with meaning. Design allows us to both respond and invent. It is driven as much by desire as by necessity.

The act of designing is carried out in many different ways, from the personal choices we make every day when we set the table or plant a garden, to the collective decisions made in the marketplace or at city hall. Because it occurs on many different levels, employing different processes and degrees of expertise, design can be understood variously as craft, style, engineering, invention, planning, refinement, as an exercise in taste or an act of choice.

As design resists definition, so do the Museum's collections. Here you can find 18th-century buttons from

Haiti alongside wallpaper designed by Frank Lloyd Wright, and a Renaissance drawing for a salt cellar by Giulio Romano next to Henry Dreyfuss's sketches for a Polaroid camera. The common bond among the chairs, the lace collars, the wallpaper borders, the plates and glasses, the telephones and the trade catalogues, the costume jewelry, and rare books of architecture is that all of these articles once lived in the world. That world may have been as confined as an artisan's workroom or a salesman's briefcase, or as cosmopolitan as a transatlantic liner or a corporate boardroom. The Museum collects the tools of design and the documents of designers' ideas, as well as the designs themselves. Whether on paper or in full-blown form, we collect ideas about living made real.

This book invites you to consider the collections of the Museum in light of their part in the landscape of daily life, in shaping the spaces in which those days are passed, and in framing our communications. The first section of the book explores how design marks the hours that take us from the breakfast table to the office, to a night on the town and evening's rest. The pleasure of morning coffee, the satisfaction of a job well done, the success of a party, and the comfort of sleep all depend on an orchestra of objects. Those cups and saucers, desks and chairs, invitations and gifts, couches and beds are the armature of our days. However, servicing our needs is only half the story, as the collections affirm. Necessity is also the catalyst for invention, for breathtaking feats of color, form, technical virtuosity, and those ineffable qualities that give deep satisfaction to their owners and account for the obsessive passion of collectors.

Our relationship to the material world is one of dependency and infatuation. In shaping the spaces that we live in both conditions must be served. The second section of this book attests to the inventive energies we have brought to bear in transforming shelters into homes, and homes into towns and cities. Drawings of steeples and staircases, elaborately crafted keys and locks, precious fragments of woven hangings, panoramic wallcoverings, and countless renderings and plans speak to the ways in which the formula of four walls and a roof has yielded castles, cottages, tract homes, country houses, apartments, and office towers, each with its own character and conceits.

Shaping space can be the most public or private of acts. Likewise, the Museum's holdings range from the preparatory sketches made by architects to explore new ideas, to their presentation drawings made to persuade prospective clients. We understand space on the scale of the skyscraper, right down to the scale of the chair—an artifact with its own anatomy of arms, legs, back, and seat.

We build our houses and fill them with the necessities and the luxuries that we feel are our due, but never in isolation. Social by nature, we have devised endless means to communicate. The book's third and final section illustrates how pervasive the impulse is to commemorate, to advertise, to instruct, and to proselytize to each other. Sometimes our stories are printed in books or on boxes, sometimes they are woven into cloth or impressed on leather. Alphabets and images are designed to tell us what kind of message is being sent:

messages as light-hearted as a "wish you were here" souvenir or as ominous as a call-to-arms poster.

Even the driest of data draws on the talents of designers, indeed, depends on those talents so we can make sense of it. Calendars, maps, and graphs chart everything from the patterns of ecosystems to a company's earnings. Communication design ranges from the urgent legibility of a stop sign to the more subtle suggestiveness of a book cover. Yet those seeming polarities between utility and invention actually coexist in every act of design. A red cross is both practical and symbolic; a pop-up book both entertains and instructs its young reader.

In a sense, everything in the collections of the National Design Museum could be said to communicate because design lives. Design begins with our corporeal selves, which we make over every day, and extends to the global body politic. In the end, these collections are irrefutable evidence of how much energy we lavish on the material world, and how much we expect of it—from the basic creature comforts of life to the essential affirmation of our humanity.

On pages 184-191 the images in this book are annotated with the name of the designer, manufacturer, country, date, medium, donor or purchase credit, acquisition number, and photo credit.

The 1949 Atomic clock by George Nelson Associates is analogous to electrons orbiting the nucleus of an atom. Its design can be seen as a response to the era's fascination with atomic energy. The absence of numbers reflects the influence of European modernist design on post-war American taste.

Design for Daily Life

We live our days surrounded by a landscape of objects. Many are so familiar, they draw no special notice; others, cherished and revered, become the subject of considerable regard. Yet, whether ordinary or extraordinary, each is designed to make a difference in the quality of our lives.

An indispensable array of props—cups, plates, chairs, lamps, and tables—follow us from home to the public theater of work, play, and worship, where they are joined by legions of tools, toys, and totems. They are the supporting cast of the days of our lives. How smooth our travels, how pleasurable our meals, how productive the task, how comfortable our rest, depends on the suitability of each object to its purpose, on the inspiration of its design.

Objects increase our potency. The finger cannot write but the pen can; the hand is a poor vessel for drinking but the cup is seamless; legs tire after a few miles that trains and planes can traverse with ease. We engineer things to extend our physical prowess, and we design them to reflect our understanding of the time and place in which we live.

Arts and Crafts vase from the Grueby Pottery, Boston, MA, c. 1905.

Just as we ascribe a place to each object in our day, each has a life of its own and a story to tell about us. The designer invests each artifact with properties purposeful and suggestive. A vase brings the garden into the house; its shape and decoration distinguish it immediately from a drinking cup or a soup pot. Furthermore, the style of the vase might tell us if it sat on a coffee table in the 1950s or graced a Victorian parlor in the 1860s. Either way, it is today's heirloom, valued for its sentimental legacy as much for its material pedigree. Laden with meaning, our possessions represent the sum of the intentions of their designers and the personal associations acquired through daily use.

The gloriously diverse collections of the National Design Museum reflect the myriad ways we have designed daily existence over the millennia. They testify to the ways in which design helps us to adapt to change and, at the same time, alleviate the sameness of the everyday.

Over the centuries, our basic needs have changed little. However, the artifacts we use to meet them are extraordinarily various, for each culture approaches the same activities differently. Eating is one of those basic requirements that has spawned a veritable universe of plates, glasses, cups, knives, forks, spoons, pots, and pans, not to mention the appliances and furniture designed for the kitchen, the dining room, and the picnic table. With these artifacts, sating the appetite has assumed the ritual significance of all acts that are central to our survival.

Sunbeam Mixmaster, c. 1950.

Our cultures are defined as much by how we eat as what we eat. The fork, for example, is a relatively recent addition to the Western table setting. In medieval Europe, eating was largely accomplished by the hand. Dining was a communal affair and several people might share a single bowl, goblet, or spoon. The introduction of the individual dining fork represented a move to more fastidious table etiquette and reflected the emerging social value of the individual. Its appearance coincided with the proliferation of elaborate designs for the ritual banquets of the Italian Renaissance, when it became one of many carefully articulated elements of the aristocratic diner's place setting. However, even as the use of the fork was becoming fairly widespread in the Italian courts, in the 16th century it was still regarded as a novelty and strange custom by foreigners traveling in Italy and did not become commonplace throughout Europe for another hundred-fifty years.

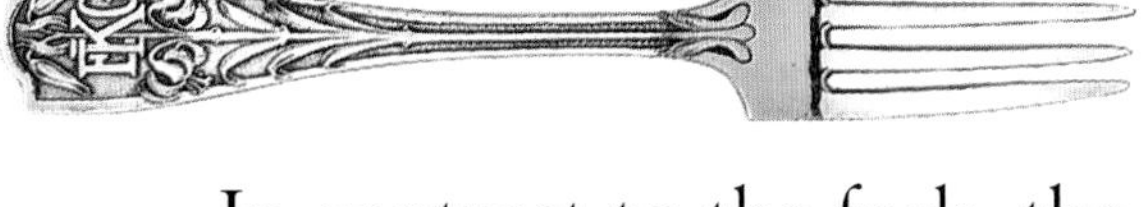

Silver fork designed by architect George Washington Maher in 1912.

In contrast to the fork, the cup has been in constant use across cultures and time, its essential form unchanged. Innovation has been largely in the realm of technical production and refinement of use.

Ancient glassblowing techniques were brought to new heights in the 15th century when the Venetians invented the highly prized and protected process of making thin, crystal-clear glass known as *cristallo*. The 19th century saw the development of mechanically pressed glass, bringing the cost of this precious commodity within reach of prosperous middle-class tables.

Whether mechanically molded or thrown on the potter's wheel, the cups, glasses, mugs, and bowls that line our shelves are each associated with different beverages and social customs, from morning coffee to high tea. Whole economies and a multitude of accessories have risen around the practice and pleasure of imbibing. The vessels of ancient bacchanalian rites have been translated into a host of specialized wares, including wine goblets, martini shakers, brandy snifters, cordial glasses, champagne flutes, beer steins, even chalices.

Gilt-bronze goblet made in honor of Andrew Carnegie in 1907.

Whether in church or in a tavern, drinking is a rite of fellowship. Designers have crafted clear forms and iconographies, often with wit and a good measure of irony, to tell us which occasion we toast. The penguin-shaped cocktail shakers that were popular in the 1930s announced the hour to unwind when work and play had been firmly segregated in the 20th-century day. A modest enamel-painted beaker from 18th-century Switzerland became a vehicle to satirize the hunt, with a rifle-toting rabbit turning the tables on the customary relationship of hare to hunter.

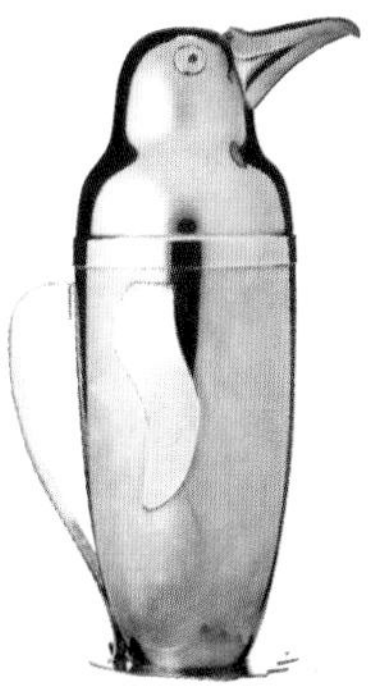

Silver-plated cocktail shaker designed by Emil A. Schuelke, 1936.

18th-century Swiss beaker.

The degree of invention and enterprise invested in design for eating and drinking reveals just how seriously we take the nourishment of the body—the most familiar object of our attentions. The impulse to protect and to fetishize our own external form reaches its apotheosis in jewelry. Like a bird's plumage, jewelry sends a signal of gender and status, drawing the eye to the body part adorned. The bracelets that decoratively jangle from our wrists today may have had their origins in armor made to protect and strengthen a vulnerable joint. The waistline has been girded by belts that held useful and handsome accessories within ready reach of the hand, from pocket watches to chatelaines—the belt hooks that held keys, money, or scents before the days of handbags. Perhaps the most powerful site of the jeweler's attentions has been the head. The center of our thoughts and senses, it has been crowned, capped, pierced, and painted in endless variety. Hair and hat ornaments are some of the most elaborate forms of jewelry in the traditional dress of many parts of the globe. In the 19th century, jewelry was actually crafted of human hair— a token of affection for a loved one, completing the circle of body and ornament.

Hand chatelaine, c. 1890.

Memorial bracelet made with hair, c. 1837

More common fibers of cotton, silk, wool, and linen make up our most necessary garb, but in the hands of the designer they become a paint box of colors and an architecture of form. Skeletal in nature, lace is the epitome of cloth as structure. It is as intricate and dimensional as molecular patterns and, likewise, requires the aid of a microscope to be fully seen in all its exquisite detail. However, the appeal of lace derives as much from its labor-intensive virtuosity as from its sheer beauty.

In 17th-century France lace was so prized—by both sexes—that exorbitant sums of money were being drained from French coffers to purchase the precious commodity abroad, prompting Louis XIV's finance minister Colbert to issue a decree in 1665 establishing a state-supported industry. From then on, laces such as the 17th-century man's collar in the Museum's collection made in Italy (seen on page 38) would be produced and purchased domestically in France.

If reverence for the body and its garb seems occasionally to border on the profane, it is also true that we have exercised our creativity no less energetically in pursuit of the sacred. In an eternal effort to reconcile the mystery of our existence, we have developed a host of religions, cults, and secular societies, each with its own deities, rites, and symbols. Some believe it sacrilegious to represent and name the Unknowable; others celebrate divinity in images and icons. The objects used in worship and meditation reveal these convictions.

An exquisitely illustrated 15th-century prayer book, or book of hours, explicitly reflects Catholicism's reverence for Mary, the mother of Christ. In all likelihood, a noblewoman structured her devotions to the "Blessed Mother" through the prayers and hand-painted pictures of holy scenes contained in this rare book.

15th-century prayer book.

By contrast, Judaism discourages images of people in its religious artifacts. Torah pointers, or *yads*, such as the 18th- and 19th-century examples in the collection, illustrate how the human is distanced from the heavenly. The Hebrew word *yad* literally means

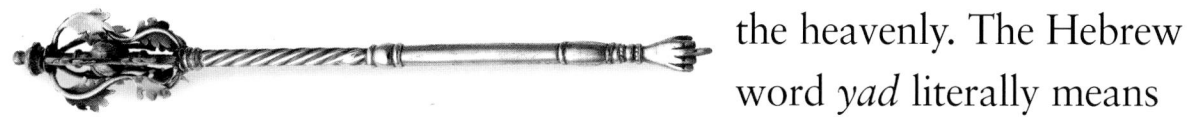

18th-century silver Torah pointer.

hand, and the pointer is used in lieu of the reader's finger to guide the eye across the Torah. The scepter-like wand respects the holiness of the sacred scroll, but also serves to protect the parchment, passed down from generation to generation.

The quest for spiritual, transcendent experiences has an earthly parallel in the wanderlust that has impelled our travels and explorations since the invention of the wheel. Designers have shaped and styled conveyances of all manner—the rickshaws, carts, carriages, cars, boats, and trains that carry us in ceremonial parades and military convoys, on holiday voyages and daily office commutes. None of these has so radically altered our access to the world as the airplane. Forever changing the flow of goods and people, air travel collapsed world destinations from distances of days or weeks into hours. A journey that might have been broken by rest stops at road houses is now compressed into a single seating. The body must adjust to the strictures of confinement at the same time it is propelled further, faster than ever before. In the 20th century, the profession of industrial design was born in large part to adapt these new mechanical appendages to our physical selves. The Museum's archives document how one of its most influential practitioners, Henry Dreyfuss, devoted much of his career to the traveler's well-being and comfort.

20th-century luggage label.

In 1955, Dreyfuss was commissioned by the Lockheed Corporation to design the Electra 188, an all new turbo-prop plane built for short flights. The Dreyfuss firm saw an opportunity to study the effects of extended seating, particularly as it would affect economy-class

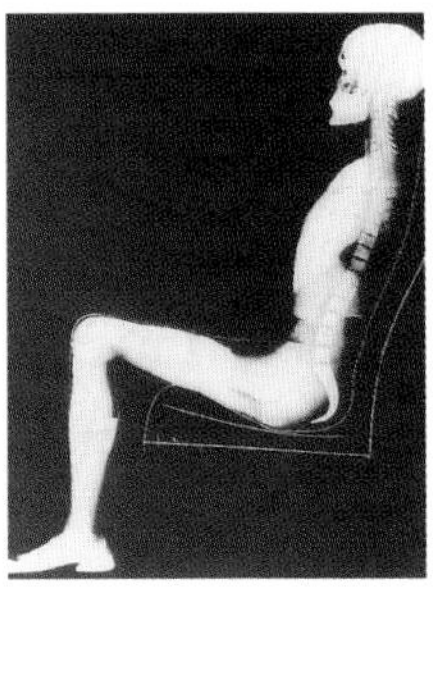
X-ray analysis from Henry Dreyfuss's report to the Lockheed Corporation, 1955.

travelers. Their report to Lockheed was based on original research conducted by the firm's human factors consultant Dr. Janet Travell, who would later become John F. Kennedy's and Lyndon B. Johnson's physician. Travell tested the firm's designs with people who would feel physical fatigue most quickly: those with muscular or skeletal disorders, her area of specialization. The design process that generated the Electra's seat design is prototypical of the comprehensive, scientific, and behavioral research that is the hallmark of successful product design today.

The Electra was designed primarily for short, quick business trips, and indeed, they are the mainstay of the airline industry. However, most of us still associate long-distance travel with the prospect of vacation. Dreyfuss's famed redesign of the 20th Century Limited train in 1938 celebrated the romance of travel from its luxuriously appointed sleeping cars down to the coordinated menus and dining accessories. The traveler was invited to relish the journey, to indulge in the escapism afforded by distance that was possible before the advent of cellular phones and faxes.

20th Century Limited dining appointments by Henry Dreyfuss, 1938.

No matter how persistently the work ethic may call us, we have always made time for leisure, creating complete environments—stadia, gyms, theaters, ballrooms, playgrounds, even special rooms in our homes—for the pursuit of pleasure. Cultural concepts of childhood can be examined through the toys and playthings each generation leaves behind. In our own time, the duration of childhood has been noticeably extended—witness the burgeoning field of sports equipment for adults or the popularity of theme parks for all ages. This phenomenon is often attributed to mid-century prosperity and the new premium put on life

Beach party wallpaper, 1958.

in the wake of World War II. Teenagers became a recognized demographic group in the 1950s, and even the traditionally conservative wallcoverings industry responded to this new market with papers patterned with cheerleaders, prom corsages, and beach parties.

Contemporary designers are less preoccupied with the myth of youth than the realities of middle age. Today's consumers, thoroughly accustomed to convenience, will insist on a world designed to accommodate their needs as they change over time. In the process, we will move closer to a model of universal design—with products and environments designed to be used through the life span by the broadest number of people. Daily life is full of tasks, chores, and small pleasures that constantly engage us with objects. Our dignity and sense of self depend on our ability to control them and to choose them to reflect our individual tastes. The challenge ahead is nothing less than the ongoing democratization of design.

1

Signs of social status infiltrate our wardrobes, reinforcing hierarchies of wealth, position, and taste.

1 In China during the Ming Dynasty (1368-1644), civil officials displayed their rank through specific insignia worn on their clothing. The crane on this badge identified the wearer as belonging to the highest rank.

2 A brightly colored bouquet of flowers adorns the front of this man's silk hat from 18th-century France or Italy. 3 A rare, painted paper fan from early 19th-century England or France depicts a scene from Greek mythology. In the center, Athena, who represents wisdom, chastity, and victory in war, receives tribute from a kneeling virgin.

2

3

"Nothing can be ascertained about Mrs. Brown, the owner of this magnificent shoe, except her name," records T. Watson Greig of Scotland. In 1885, seeing his outstanding collection of shoes age and crumble, he commissioned a book to document the lot, to which he added meticulous notes about the former occupants.

1

2

About these shoes, Greig, Vice President of the Literary and Antiquarian Society of Perthshire, notes: 1 "Mrs. Brown's shoes date about the time of Queen Elizabeth." 2 "This shoe belonged to Lilias, daughter of the 12th Earl of Eglinton, and was worn by her at her marriage about the middle of the 18th century." 3 "From its appearance this shoe must be a very old one, but the date and name of its wearer cannot be discovered."

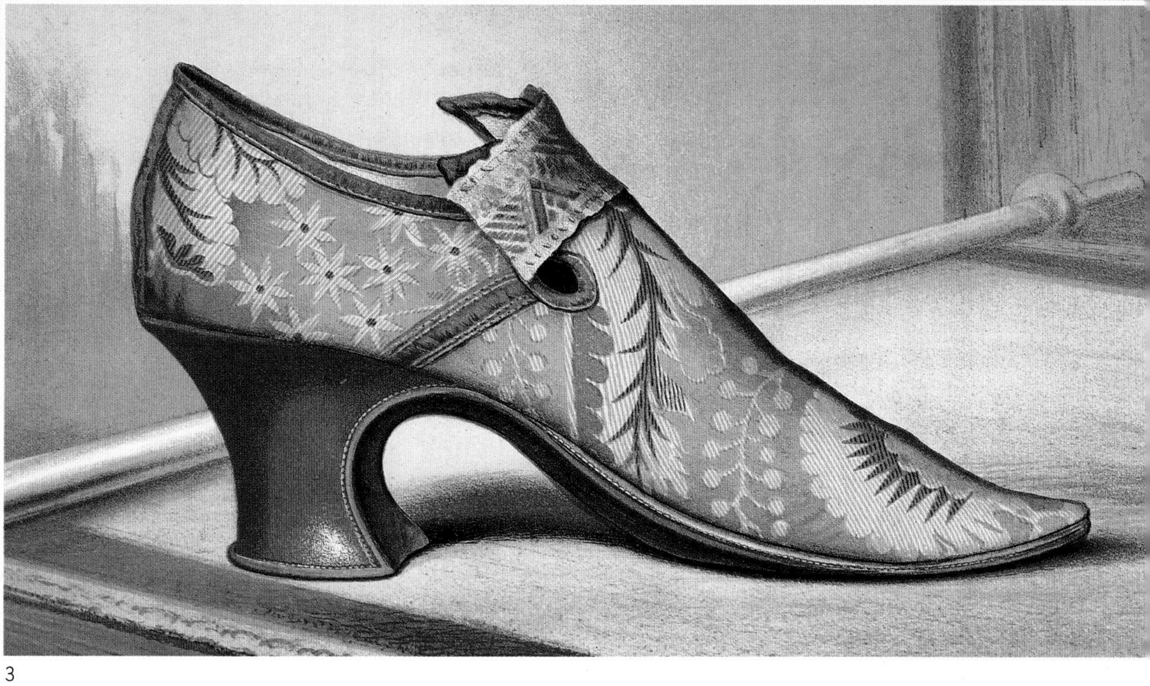

3

Do clothes make the man? They do send signals about the wearer and the occasion, whether in 19th-century England or 20th-century Zaire.

1 Plate from *The Cyclopaedia of the British Costumes from the Metropolitan Repository of Fashions,* 1843. 2 This piece of a man's ceremonial dance skirt from the Ngende or Ngongo tribe of the Kuba Kingdom in Zaire was made in the 20th century. Decorated with an unusual resist dyeing technique, the fabric was folded and stitched tightly through all layers, creating areas reserved from the dye.

1

2

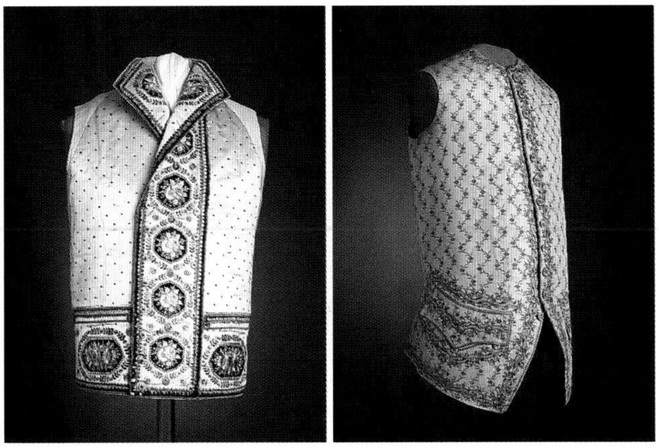

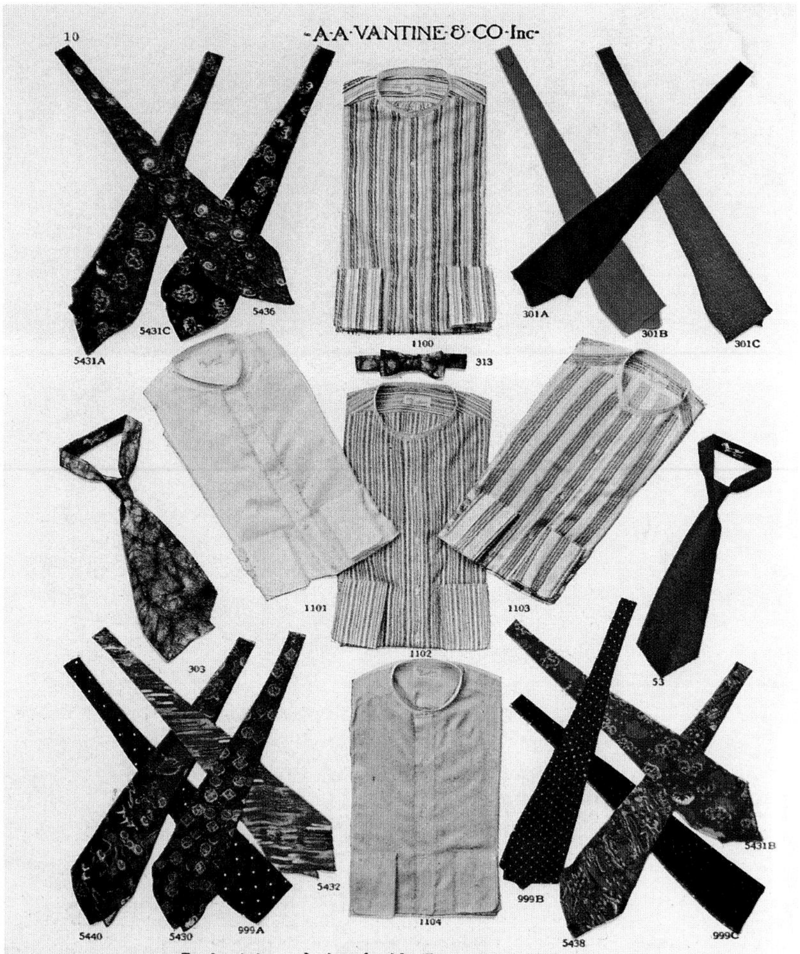

1

2

3

1 This colorful 20th-century walking stick by Fred Brown features portraits of early film actresses Mary Pickford, Clara Young, Mary Miles Minter, and Bebe Daniels.

2 Men's neckwear and shirts from the 1918 catalogue designed for the "out-of-town patrons" of A. A. Vantine's "The Oriental Store" in New York. 3 Waistcoats, such as those shown here, were an elaborately decorated part of an 18th-century French gentleman's formal attire, worn with a coat and breeches, white stockings, and buckled shoes.

Masculine attire has never been exempt from ornament, from embroidered waistcoats and flashy walking sticks to the latest shirts and ties.

2

2 Dagobert Peche's harlequin-inspired dress design from 1914-15 exemplifies the charm and caprice of Wiener Werkstätte fashions, which emphasized distinctive, unusual patterns over practicality and wearability.

1

Pattern plays in counterpoint to the lines of the body and its garb.

1 A 1919 watercolor and gouache study shows color variations for the textile design "Diomedes," by Dagobert Peche, director of the Austrian Wiener Werkstätte design studio from 1919 to 1923. In Wiener Werkstätte textiles, simplified geometric forms prevailed over the ornate floral and vegetal patterns of the Art Nouveau style.

3 Innovative Japanese textile designer Junichi Arai used three different kinds of nylon yarn to create "The Big Checkerboard" in 1984, seen here in detail.

Opposite, far right: 4 This embroidered and appliquéd cotton robe was made by a member of the Ainu people of Hokkaido, Japan, late in the 19th century, or early in the 20th. The bold design, shown in detail, identifies the Ainu culture.

3

4

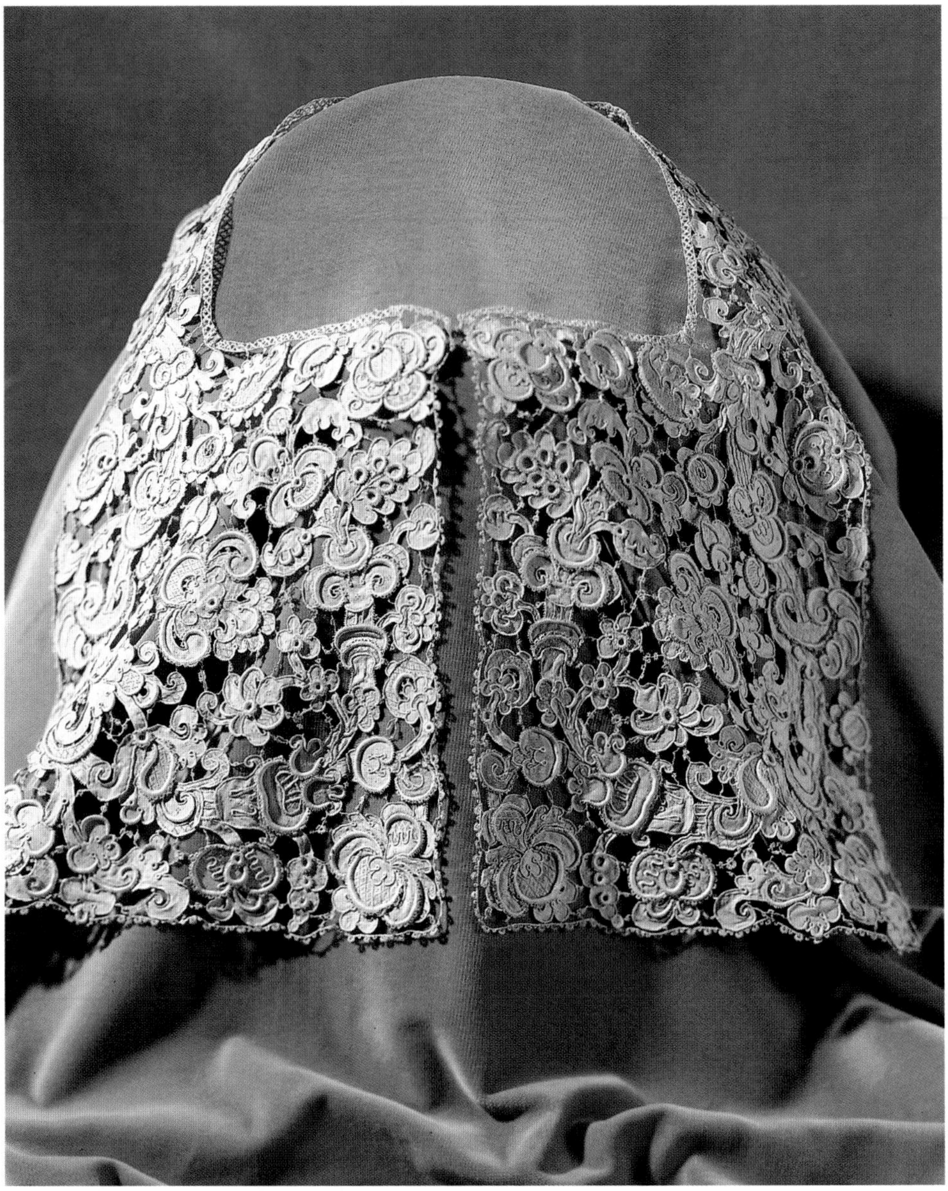

1

Finery has always pushed its material limits, from the delicate threads of the finest lace and embroidery to the novel elegance of intricately sculpted jewelry.

1 In the 17th century, alarmed that enormous amounts of money were leaving France to purchase lace, Louis XIV's finance minister Colbert issued a proclamation establishing a state-supported lace industry. This man's collar is one of the imported pieces that from then on would be produced in France. It is made of an Italian lace, commonly called Venetian, characterized by bold motifs with thick, raised edges and delicate connecting bars. 2 This imaginatively embroidered hat would have been worn by an Englishman in about 1600. 3 This suite of jewelry, made in the 1850s and thought to be French, is unusual for its use of engraved aluminum in gold settings. Aluminum was considered a rare and expensive metal until the 1890s, when it was first processed in large quantities. Queen Victoria of England and Empress Eugénie of France both owned jewelry made with aluminum.

2

1

2

3

4

1 Gold and silver threads were used for the elaborate knotting of this purse from early 17th-century England. 2 Art Deco German necklace made of silver, moonstone, and marcasite, c. 1930s. 3 This necklace from Kenya has three shell ornaments that hang from a beaded leather strap, indicating that the wearer had three sons who were warriors. 4 Art Nouveau belt buckle made of gold, opal, pearls, and garnets, attributed to Edward Colonna, c. 1900. 5 18th-century scent containers. 6 A Chinese hair pin, c. 1890, with carved jade plaques and gold-work dragons with gem-set eyes.

5

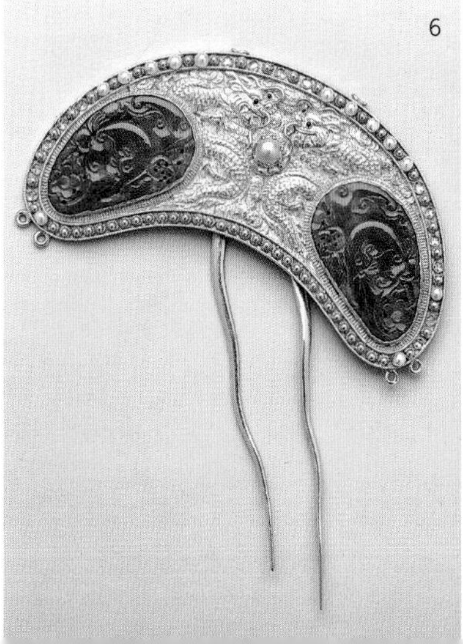
6

7

8

Baubles, bangles, beads, and even Goodyear rubber are called into the service of fashion.

7 At the end of the 19th century, chatelaines were worn on the wrist as a useful form of jewelry adorned with practical attachments—in this case, a sewing kit, notepad, locket, eyeglasses, purse, scent case, and mirror. 8 In 1844, Charles Goodyear invented the process of vulcanization, revolutionizing the industrial potential of rubber. This fan (1845-60) showed that vulcanized rubber could be formed in thin, yet strong shapes. Used to cool the face, it proved that the smell of rubber had been removed through Goodyear's new process.

9

9 This 19th-century purse is an example of needle-made netting, known as *oya*, done in the Eastern Mediterranean. 10 Diamond brooch, England or France, c. 1800.

10

1

1 A parure, or matching suite of jewelry, c. 1810, with micro-mosaic plaques showing famous Roman architectural sites including the Coliseum, shown in detail. Each plaque is made from hundreds of tiny chips of hardstones. 2 This French salesman's sample book from the late 18th century offered customers a variety of buttons to choose from when ordering a matching set for a coat or other article of clothing.

A world of ornament is enframed by a geometry of ellipses, circles, and ovals.

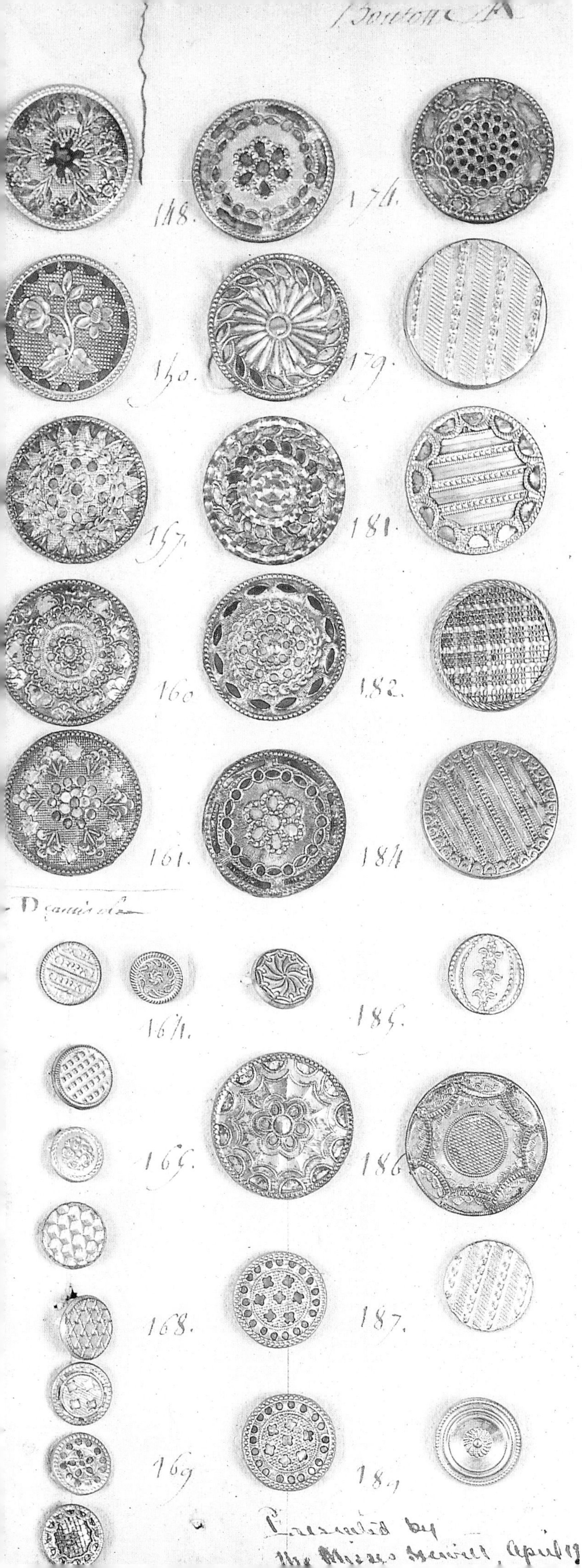

2

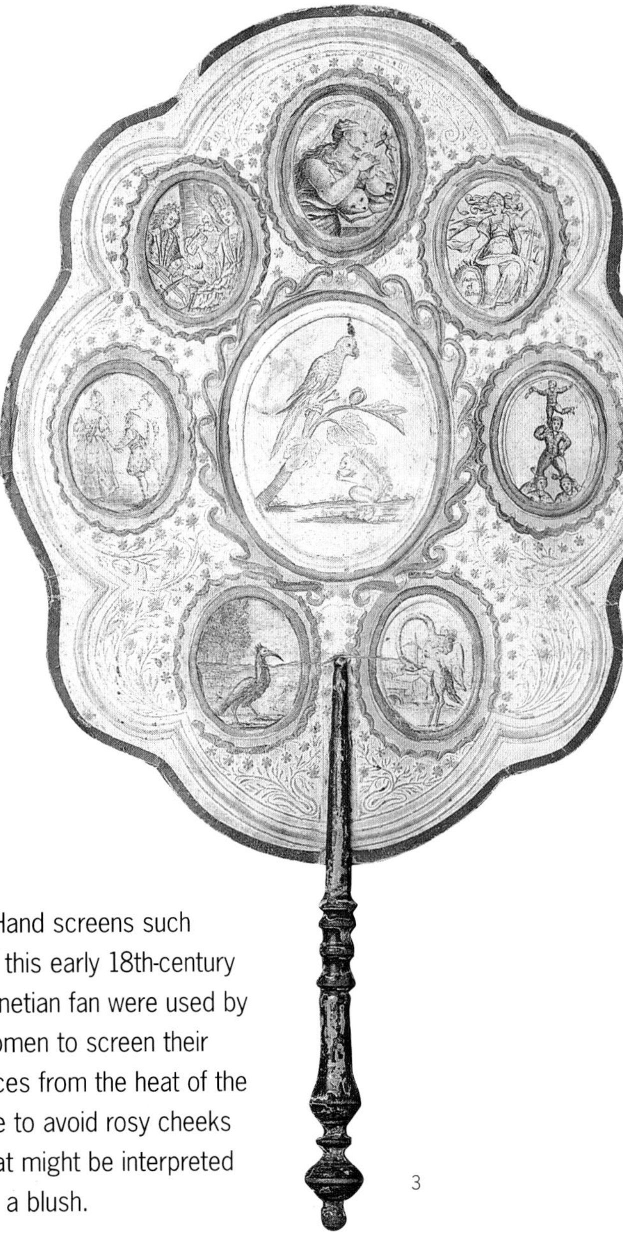

3

3 Hand screens such as this early 18th-century Venetian fan were used by women to screen their faces from the heat of the fire to avoid rosy cheeks that might be interpreted as a blush.

1 These practical plumbing fixtures, illustrated in an 1888 trade catalogue of the J.L. Mott Iron Works, were advertised as both "artistic and beautiful to look upon." 2 An illustration of toilet ceramics from the 1889 catalogue of *La Maison du Grand Dépot* in Paris. Their decorative motifs reflect the popularity of the *japonisme* style at the turn of the century.

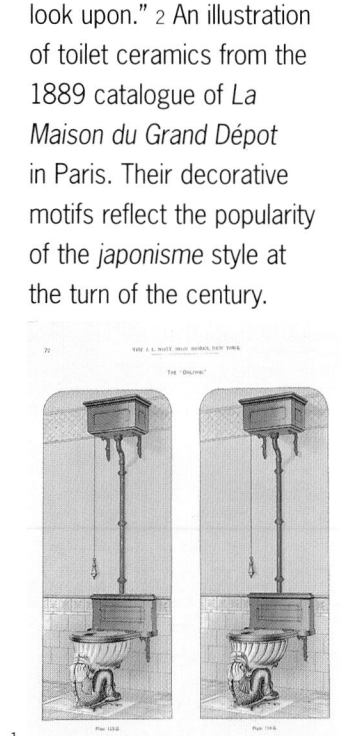

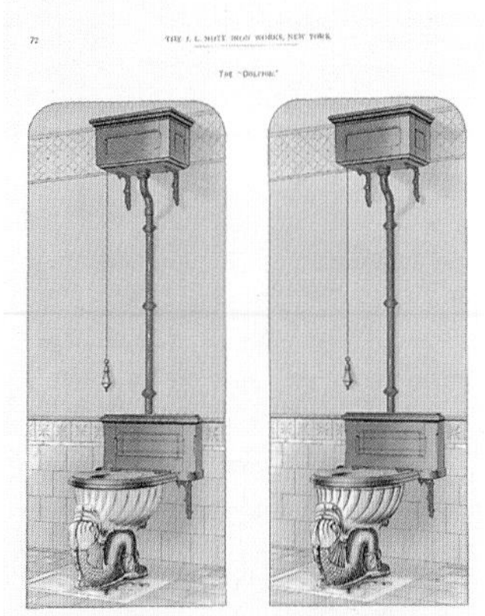

1

2

3

3 Photograph of the Antoine beauty salon in Paris, c. 1929. 4 *The Children's Bath*, engraved by Israhel van Meckenem, gives an unusually intimate glimpse of 15th-century domestic life.

4

5

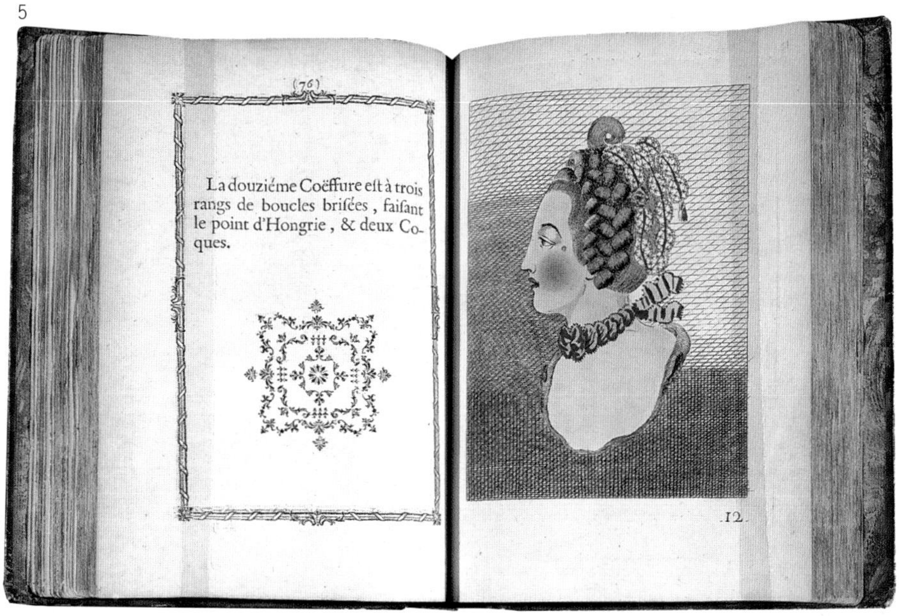
(76)
La douziéme Coëffure eſt à trois rangs de boucles briſées, faiſant le point d'Hongrie, & deux Coques.
12.

5 *L'Art de la Coëffure...*, an 18th-century hair styling manual produced during the reign of Louis XV, served as an indispensable guide to the era's elaborate coiffures.

In the name of vanity, even the most mundane needs are subject to intense design scrutiny.

6

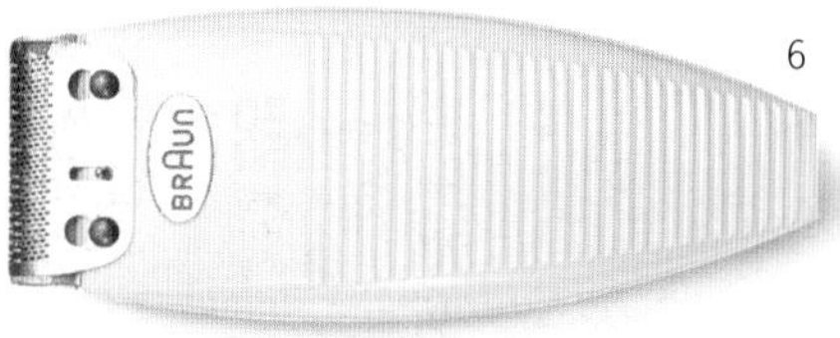

7

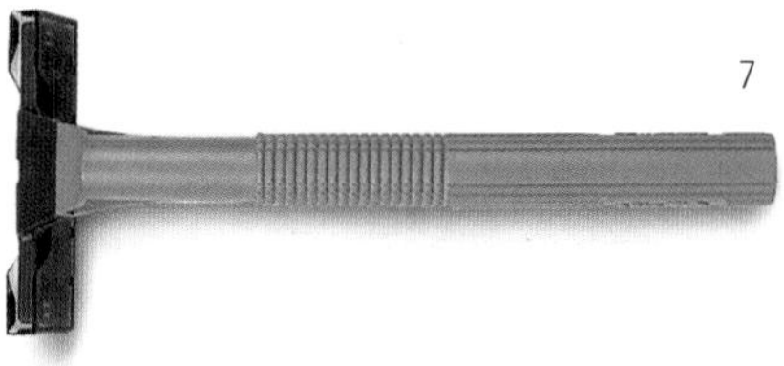

7

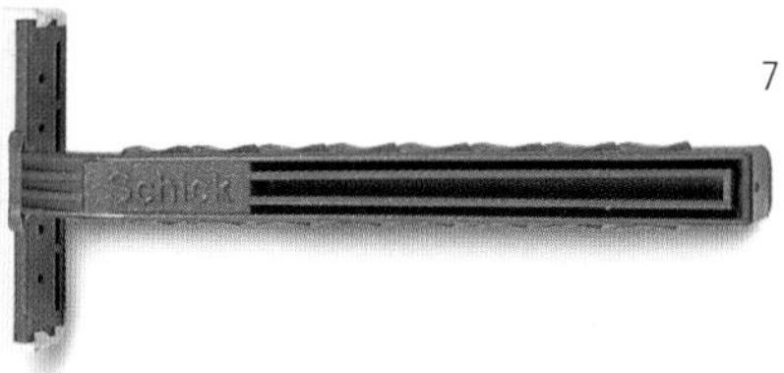

8

6 Braun's S50 Standard Electric Shaver from 1950. 7 Disposable razors by Gillette and Schick from the 1980s. 8 The 1992 Wally razor created by Hoke 2 features a wide, non-slip handle and a suction holder that allows it to be attached to the wall of a shower. 9 Introduced in 1983, the Radius toothbrush, designed by James O'Halloran, features an extra-width handle that is easy to hold and comes in both right- and left-handed models.

9

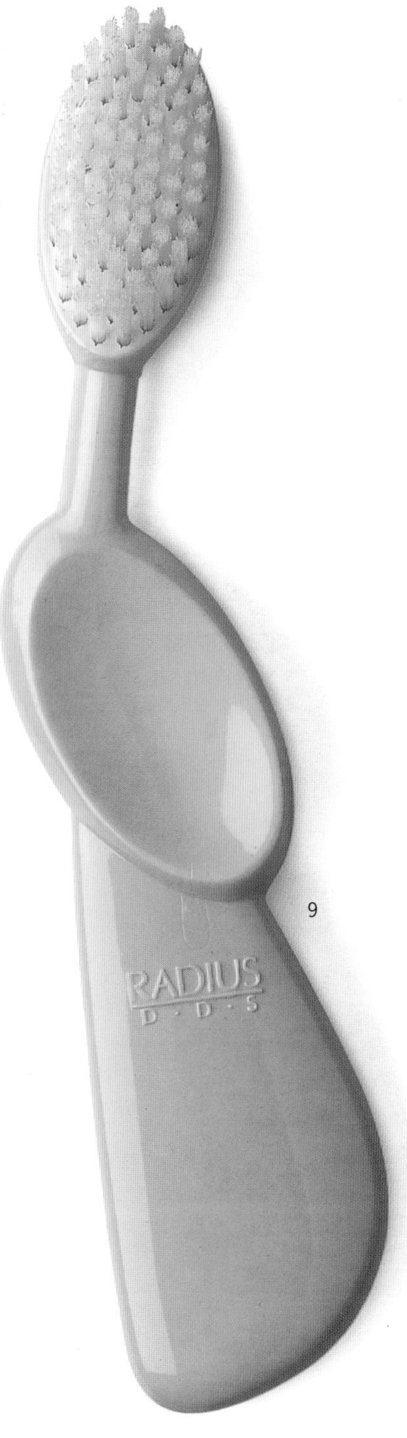

1

The spiritual is expressed by gestures in both artifacts and architecture.

2

1 This temple hanging was made in Tibet in the 18th or 19th century of earlier Chinese fabrics of various types and dates. The standing Buddhist spiritual figure, or *bodhisattva,* is a remarkable and rare needlework, dating to the Chinese Yuan period (1279-1368).

2 Architect and painter Charles de Wailly's 1789 watercolor offers a dramatic view of the marble pulpit he designed for the Church of Saint Sulpice in Paris.

1

1 Lions are ancient symbols that have been used to protect temples, tombs, and palaces in the East and West for millennia. This part-lion, part-female wooden figure was designed as an architectural corner element for a Buddhist temple in Thailand, possibly in the 18th century. 2 A *yad* is a pointer used by the reader of the Torah to follow the text. *Yads* can range in design from simple wooden posts to elaborately ornamented rods, such as this 18th-century silver example from Italy.

2

The sacred is translated through earthly media as rich and diverse as the religions of the world.

3

3 This early 17th-century Portuguese lace border illustrates the biblical story in which Judith severs Holofernes' head to save her people. 4 Silk weaving flourished in Spain under Muslim rule. This late 14th- or early 15th-century example, bearing an Arabic inscription praising the Sultan, was adapted for use as an ecclesiastical vestment of the Catholic church when it was cut into the shape of a hood for a cope. 5 This 14th-century Italian design, showing a figure in the jaws of a fish and the head of Christ surrounded by seraphim, is the earliest drawing in the Museum's collection. It is one of only 12 in the world that, based on their similarity, are believed to be ornament designs for late 14th-century textiles. This very rare document offers a clue to the origins of European ornament.

5

4

1

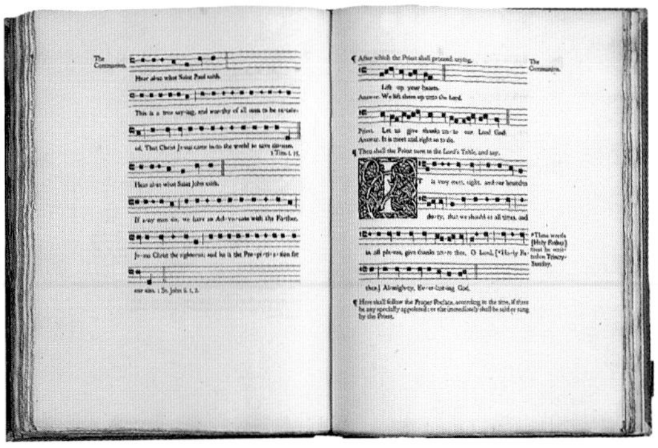

1 This 1896 altar book, commissioned by the Episcopal Church, features designs by architect Bertram Grosvenor Goodhue and others.

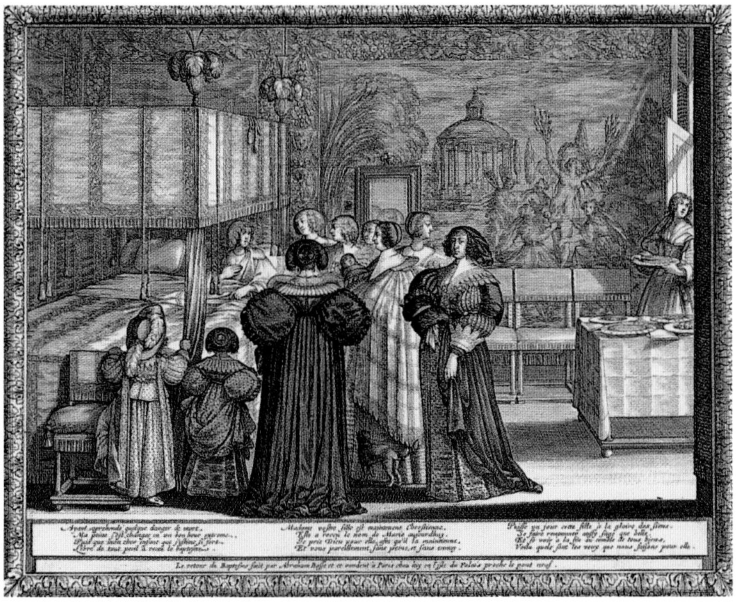

3

4

5

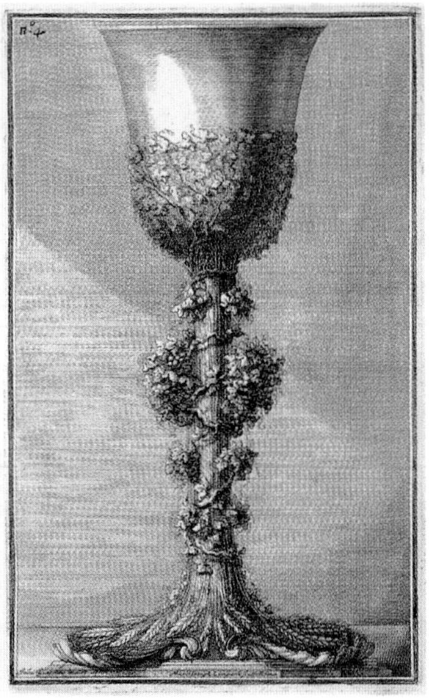

2

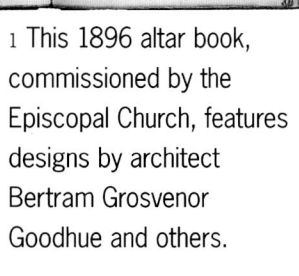

2 Scattered among the flowers in this 17th-century Spanish silk are the emblems of the Passion of Christ: the cross, cock, whip, ladder, sponge on a rod, crown of thorns, and nails. 3 Unlike today, when bedrooms are considered private domestic spaces, in 17th-century France bedchambers were often settings for public occasions or rituals, like this post-baptismal celebration.

4 This bronze 18th-century censer is typical of those used to burn incense in Christian rituals since medieval times. Their perforated surfaces allow fragrant smoke to escape, and long chains are used to swing them through the air to spread the scent. 5 A 1714 engraving of a design for a chalice by Maximilian Joseph Limpach, after Giovanni Giardini, a leading goldsmith of the baroque period.

All the senses are employed in transforming the secular to the sanctified: the touch of holy water, the smell of incense, the taste of a holy meal, the sight of symbols, and sound of voices raised in song.

6

7

6 During the depression of the 1930s, the Works Progress Administration sponsored an art course for children at the Educational Alliance on New York's Lower East Side where A. Nedby, aged 10, designed and screen-printed this cotton.

7 Priests in the late 16th or early 17th century would have worn an alb such as this one, embroidered and trimmed with bobbin lace, to conduct a Christian church service.

1

2

The act of eating, elevated to the art of dining.

1 This Greek kylix is a drinking vessel dating from the 6th century B.C.
2 An engraving from 1705 of a royal banquet on the occasion of the inauguration of Holy Roman Emperor Joseph I as Archduke of Lower Austria. 3 The flowing, curvaceous shapes of these silver candelabra made by Claude Ballin the Younger reflect the prevailing taste for the rococo style in Paris in the 1730s.

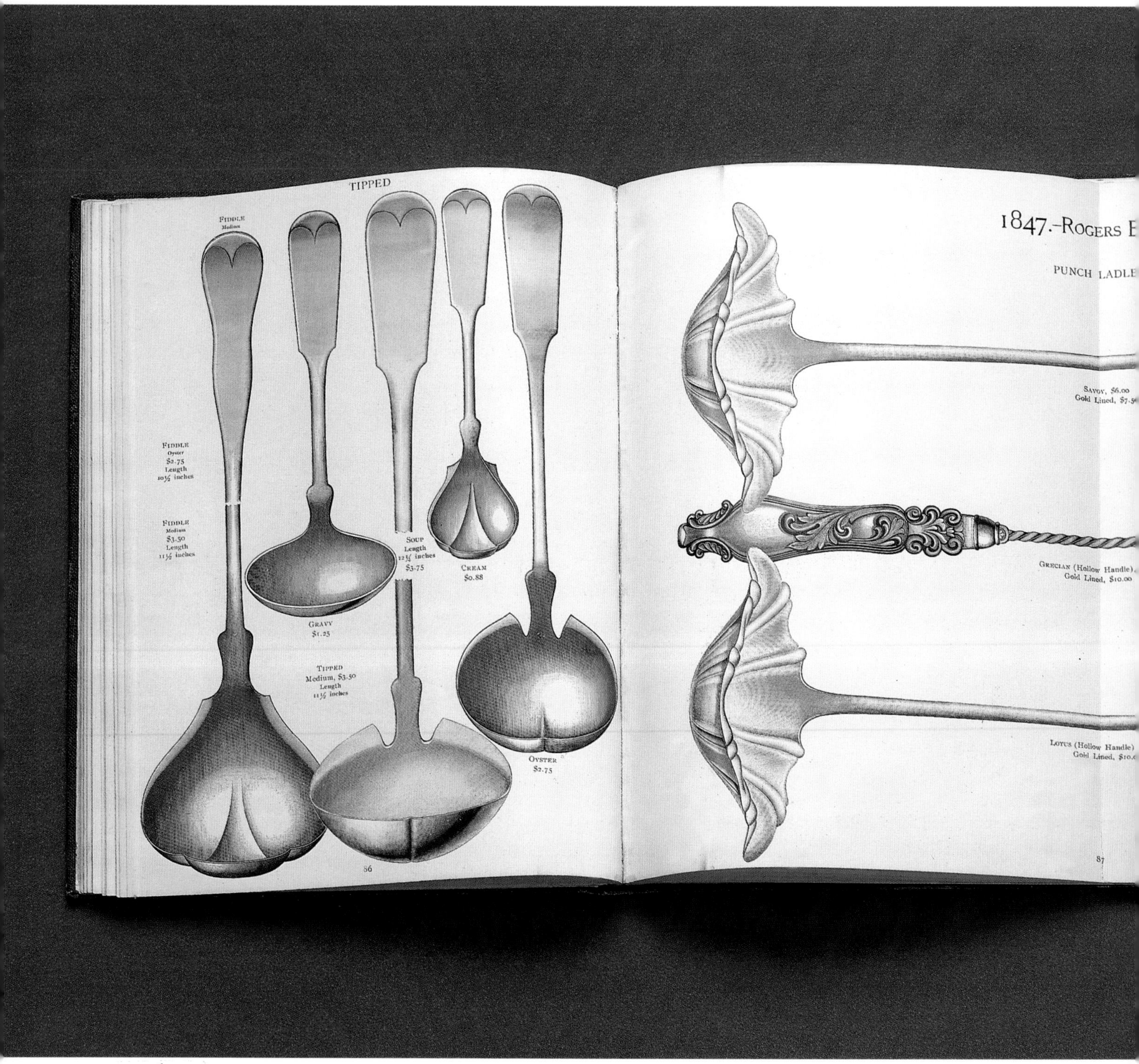

1

Form follows function… from the table to the refrigerator.

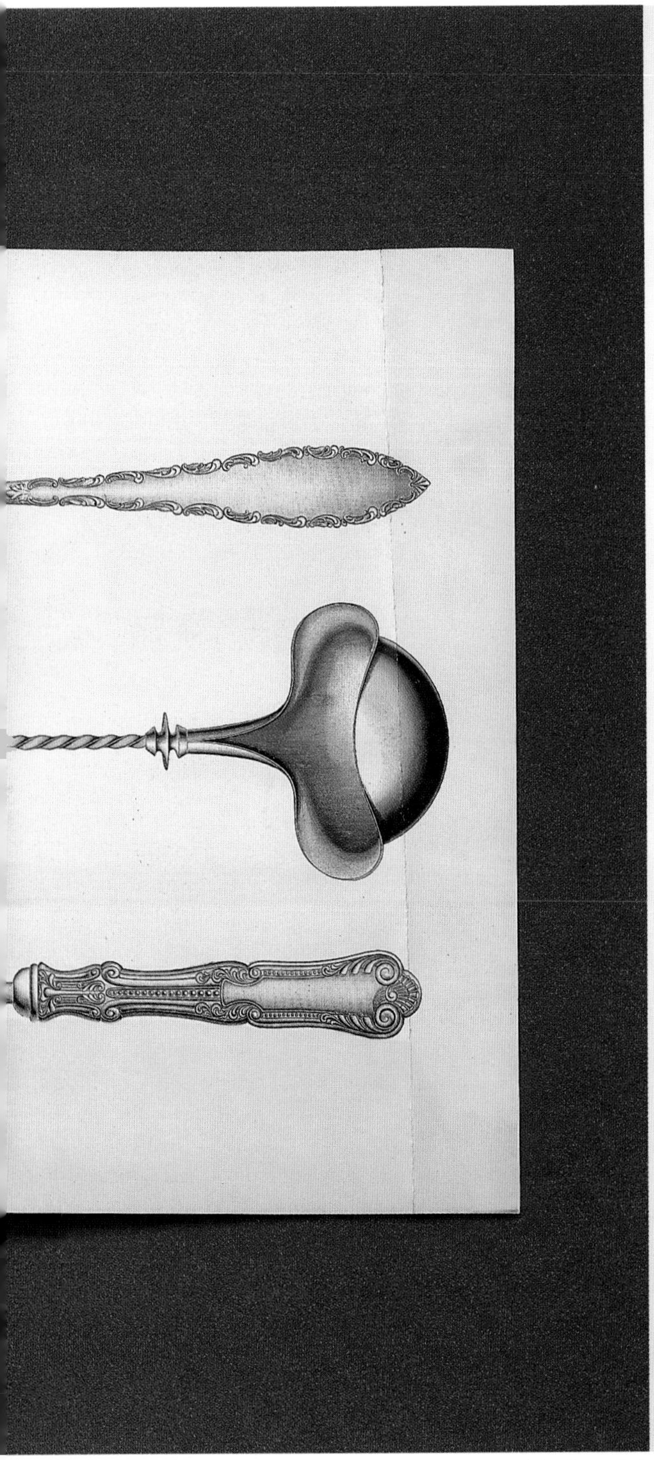

2

1 Trade catalogue illustration of silver- and gold-plated punch ladles made under the Rogers Bros. trademark in Connecticut in the late 19th century. 2 The Kubus food storage system was designed in Germany in 1938 by Wilhelm Wagenfeld for use in newly available electric refrigerators. Formed of molded glass, its containers fit compactly together to maximize the limited interior space of a refrigerator. Their geometric shapes and unornamented surfaces were easier to clean and store and reflected the modernist style of the time.

1

2

3

4

5

6

7

1 Copeland & Garret, Spode, Staffordshire, England, c. 1830. 2 "Canton" plate, Edward Colonna, Gérard, Dufraissex and Abbot, Limoges, France, 1899-1900. 3 Gabriel Pasadena, United States, 1950. 4 Pillivuyt & Cie, France, c. 1875. 5 Produced by Lavenia, Italy, 1925-30. 6 Sèvres Porcelain Factory, France, 1822. 7 Produced by Rörstrand, Sweden, mid-19th century. 8 "Travel" plate, Eric Ravilious, Wedgwood, Staffordshire, England, 1938. 9 Simon Lissim, with decoration by L. Rodzianko, France, 1927. 10 Jutta Sika, Koloman Moser School, Josef Böck Porcelain Factory, Austria, 1901-02. 11 "Homemaker" plate, Ridgeway Potteries, Staffordshire, England, c. 1955. 12 Richard Riemerschmidt, Meissen Porcelain Factory, Germany, 1903-05. 13 Frederick Hurten Rhead, "Fiesta" and "Harlequin" tablewares, Homer Laughlin China Company, United States, 1940s-1950s. 14 Jean Luce, with decoration by Charles Ahrenfeldt, France, 1937. 15 Plastic Quatrefoil centerpiece designed by Massimo Vignelli for Heller Designs, c. 1970. 16 Schramberg Pottery, Germany, c. 1928-30. 17 Kornilov Brothers, St. Petersburg, Russia, 1900-1910. 18 *"Les Pâtisseries,"* Sèvres Porcelain Factory, France, 1831. 19 The Greek A Factory, Delft, The Netherlands, 1686-1701. 20 Peter Behrens, Bauscher Brothers Porcelain Factory, Germany, c. 1901.

8

With beautiful plates, the feast begins before the meal is served.

9

10

11

12

13

14
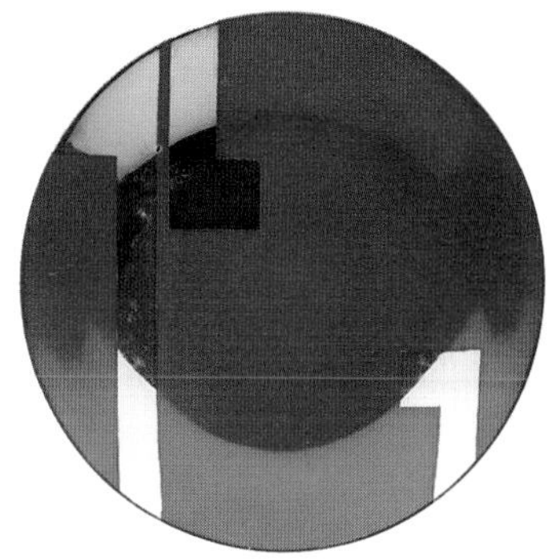

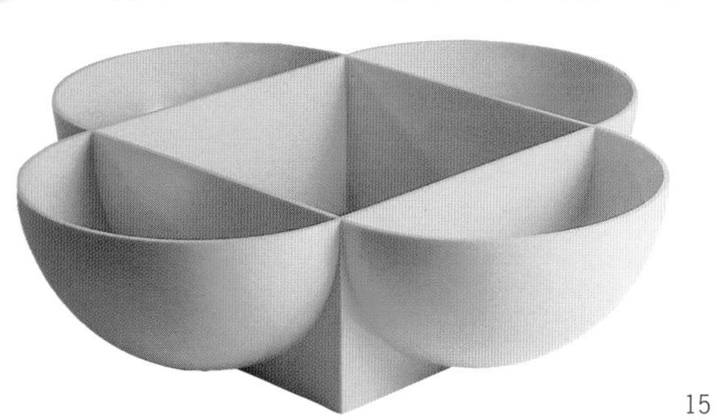
15

18

16

17

19
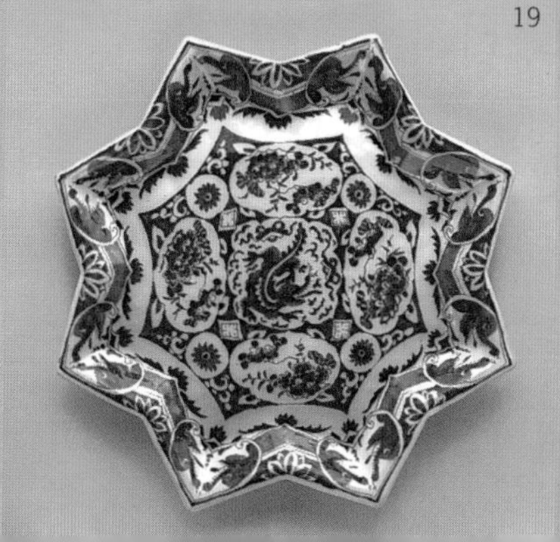

20
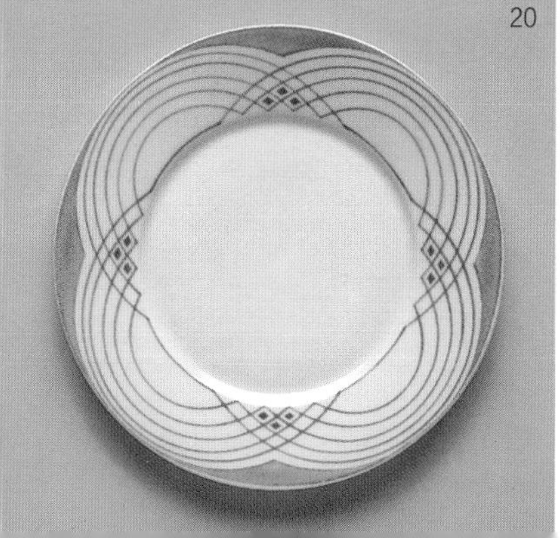

Modern masterpieces for the masses.

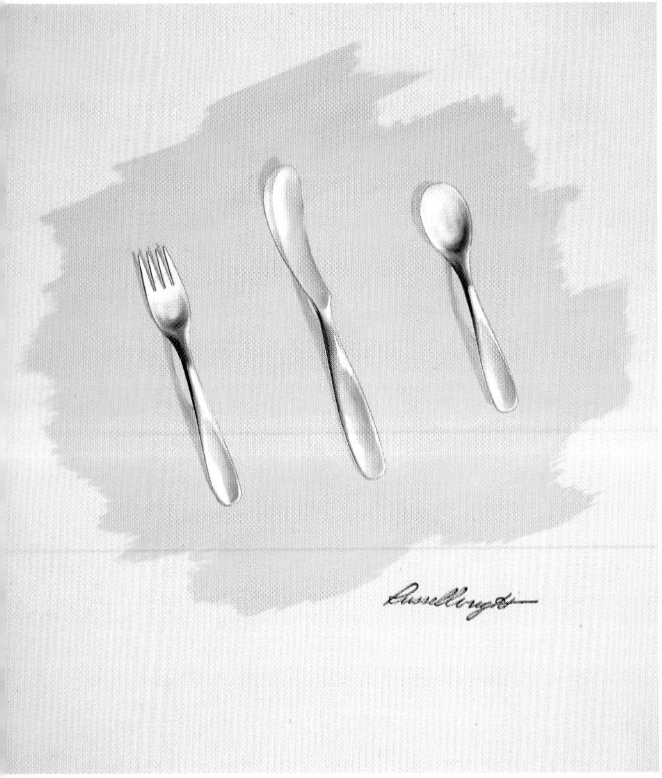

1

2

1 A 1950 drawing for stainless steel flatware by Russel Wright, who championed this new easy-care flatware as appropriate for the uniquely informal American lifestyle.

2 The "Museum" pattern tablewares designed by Eva Zeisel were exhibited at the Museum of Modern Art when they were introduced in 1946.

User-friendly design puts control in our hands.

1

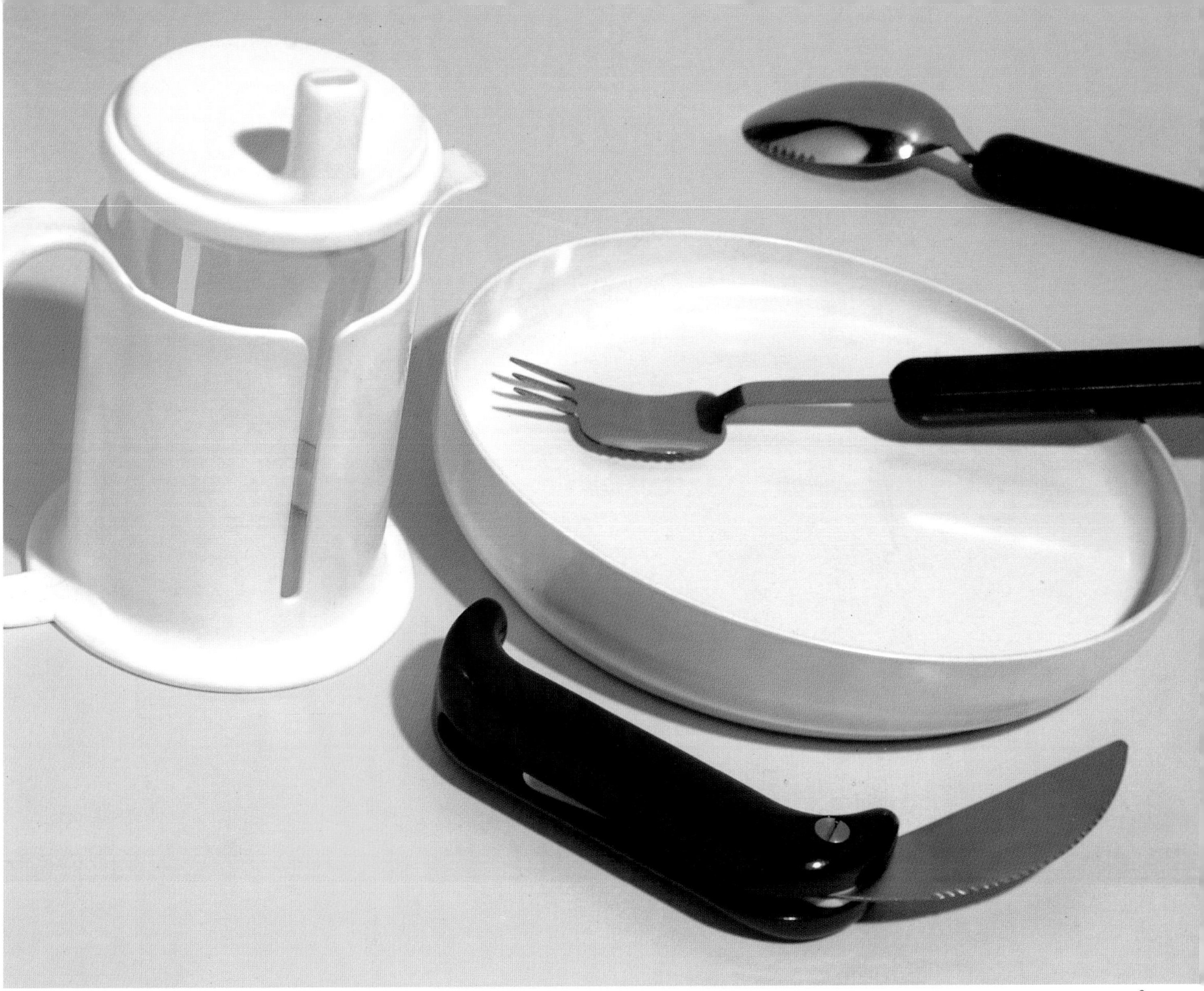

2

3

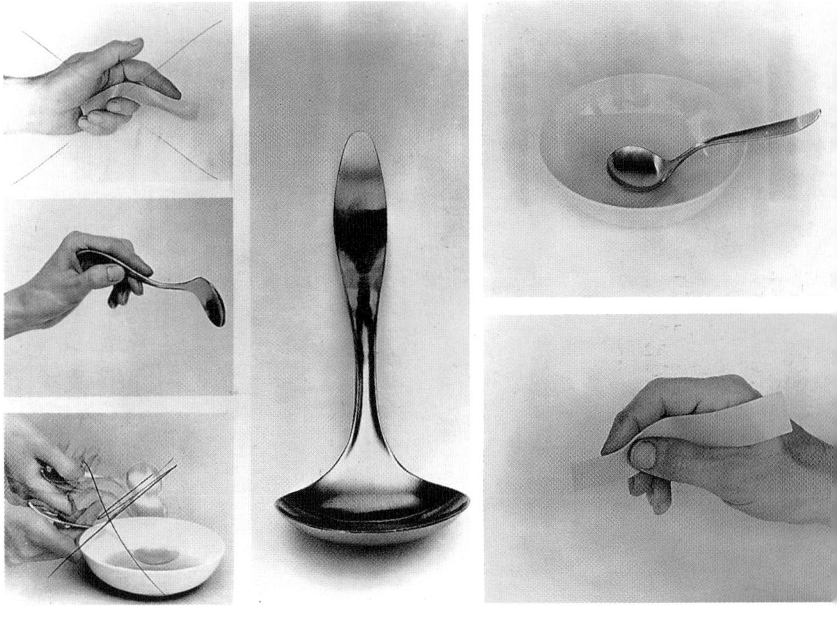

1 Designed by Smart Design to reduce hand strain in the kitchen, the soft, wide, easy-to-hold handles of the Good Grips utensils from 1991 are made of a thermoplastic rubber and are scored with flexible ridges to improve grip control. Because they take into account different users' hand strengths, Good Grips are models of universal design.

2 The Swedish firm Ergonomi Design Gruppen created this tableware in 1980 with universal design principals in mind. The set includes multipurpose utensils with extra long and wide grips, a plate with a skid-resistant bottom, and a non-spillable drinking cup. 3 These photographs show industrial designer Don Wallance's development process and the ergonomic features of the "Design One" stainless steel flatware that he created for H. E. Lauffer in 1953.

Simple tools are brought to new heights by design.

1

2

3

1 Hand-blown *Calici Natale* goblets by Carlo Moretti, 1990. 2 A soup tureen design by Louis Süe and André Mare from 1921. 3 This gilt-bronze goblet was given to Andrew Carnegie by the Engineers Club of New York in 1907. Its thistle form refers to Carnegie's Scottish heritage. Carnegie's New York residence is now the home of the National Design Museum. 4 This silver flatware was designed by architect George Washington Maher in 1912 for Rockledge, the Prairie-style summer home of Ernest and Grace King near Homer, Minnesota. The lilies in the relief decoration echo the wild flowers around Rockledge that were used as a unifying ornamental motif throughout the house.

EKG

High Spirits

1 Cocktails became widely popular in the 1920s and '30s. Since many drinks are mixed with ice, this silver-plated penguin shaker from 1936, designed by Emil A. Schuelke, would have been right at home among the new accessories created for the cocktail hour. 2 *The Savoy Cocktail Book* was compiled in 1930 by Harry Craddock of the Savoy Hotel in London as an encyclopedic guide to "every cocktail known."

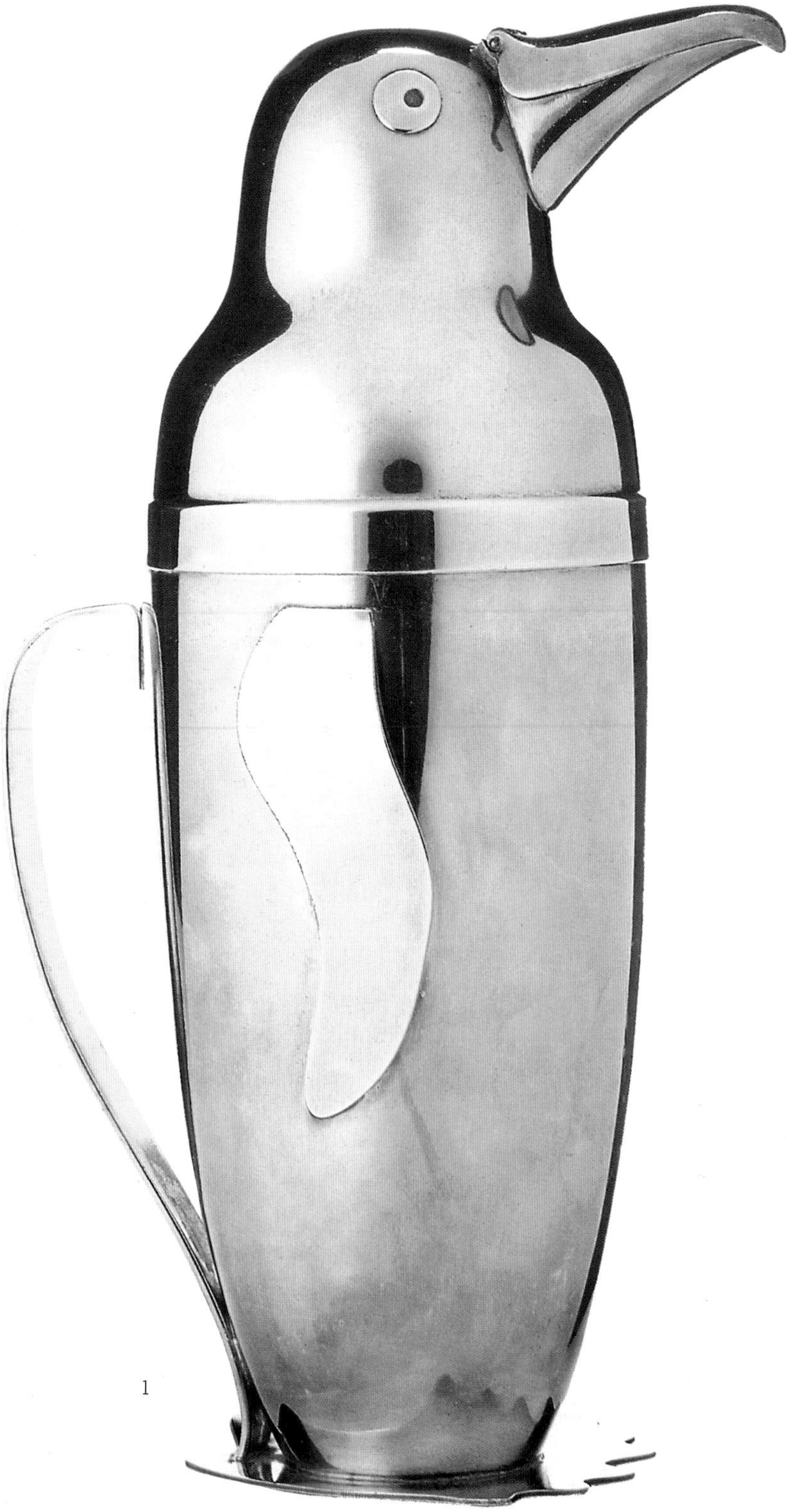

1

2

3 Photograph of bartender at Le Grand Ecart, a Paris supper club, c. 1925. The image was originally used to advertise the jewelry designed by Gérard Sandoz, shown on the model.

3

Cheers!

The basic form of the goblet—a bowl supported by a stem and foot—has been reinterpreted time and again with different treatments of color, shape, and decoration that often indicate the occasion and beverage for which the glass was designed.
1 Sherry glass, Austria, c. 1925 2 Cut-glass goblet, Irish, early 19th century 3 "Embassy" champagne glass, Walter Dorwin Teague, 1939 4 "Arkipelago" cordial glass, Timo Sarpaneva, 1979 5 "American Modern" goblet, Russel Wright, c. 1950 6 Tiffany Studios goblet, c. 1900 7 18th-century English goblet with opaque twist stem decoration 8 Etched Art Nouveau goblet, French or Belgian, c. 1900.

6
8
7
4
5

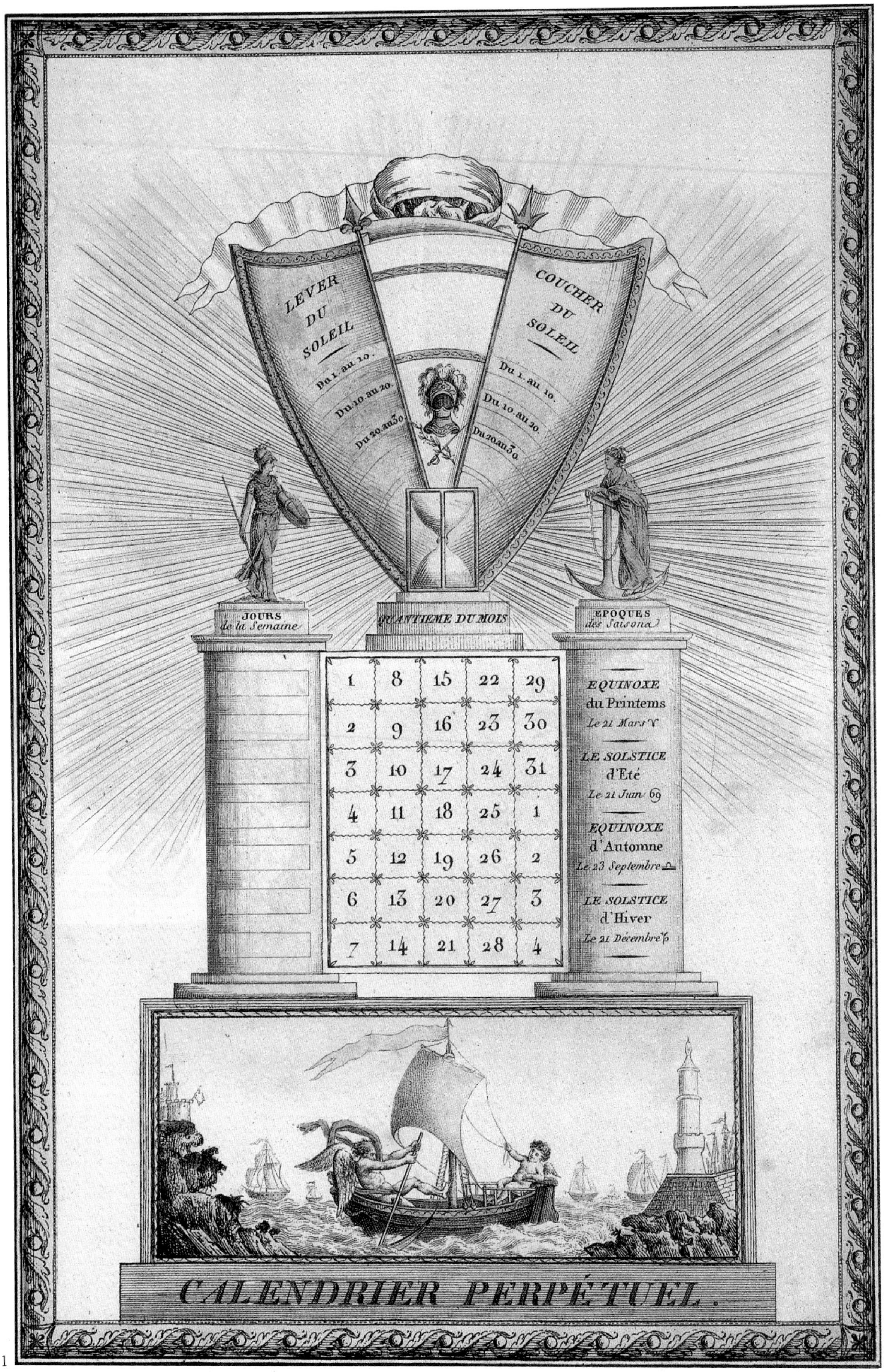
LEVER DU SOLEIL
Du 1. au 10.
Du 10. au 20.
Du 20. au 30.
COUCHER DU SOLEIL
Du 1. au 10.
Du 10. au 20.
Du 20. au 30.
JOURS de la Semaine
QUANTIEME DU MOIS
EPOQUES des Saisons
1 8 15 22 29
2 9 16 23 30
3 10 17 24 31
4 11 18 25 1
5 12 19 26 2
6 13 20 27 3
7 14 21 28 4
EQUINOXE du Printems Le 21 Mars
LE SOLSTICE d'Eté Le 21 Juin
EQUINOXE d'Automne Le 23 Septembre
LE SOLSTICE d'Hiver Le 21 Décembre
CALENDRIER PERPÉTUEL.

1

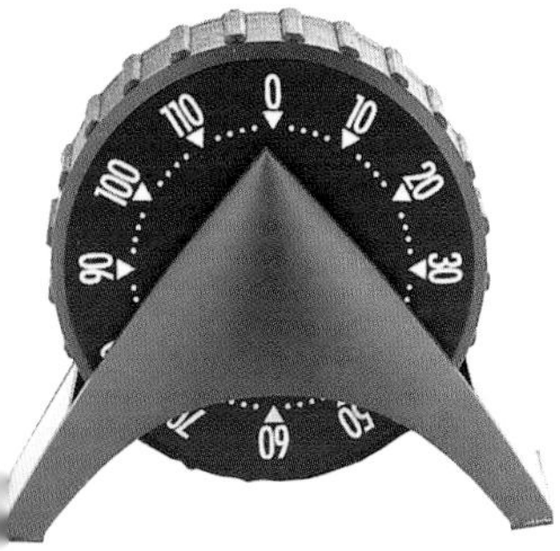
2

3

4

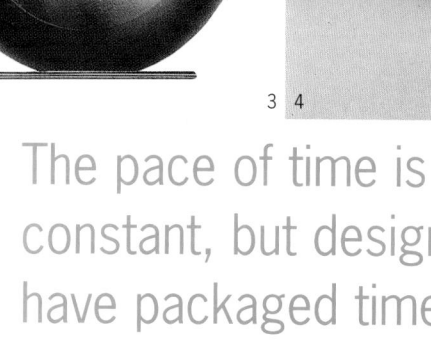

The pace of time is constant, but designers have packaged time in many measures to suit the style of the day.

5

6

1 Perpetual calendars make it possible to find the correct day of the week over a span of years. This 18th-century French calendar shows a monument with Chronos, or Father Time, at its base, navigating the ocean of life. 2 Kitchen timer by Danish designers Mads Odgård and Bernt Nobert, 1992.

3 George Nathan Horwitt, a designer best known for the Movado Museum watch, also developed digital timepieces such as the Cyclometer clock from the 1930s, shown here in a drawing. 4 The streamlined Zephyr clock, designed by Kem Weber around 1930, made early use of a digital readout panel. 5 This early 19th-century Belgian water-color design for a table clock shows an allegorical figure, possibly Hope, atop a rock encasing the clock face. At the base is a frieze with Triton and his father Neptune, who, when angered, could cause shipwrecks. 6 The Two-Time watch, Tibor Kalman, M&Co., 1980s.

"Home" work has its own archeology of artifacts and rituals, some obsolete, others eternal.

1

1 *My Lady's Chamber* was the frontispiece created by Walter Crane for the 1878 book *The House Beautiful,* which provided advice on home management, interior decoration, and etiquette. 2 The 1930 trade catalogue from the Dosco Needle Company in Germany contains samples, crochet hooks, and sewing needles used by traveling salespeople to advertise their wares. 3 Darning samplers such as this one made in the Netherlands in 1735 were used to train young women in the art of mending.

3

2
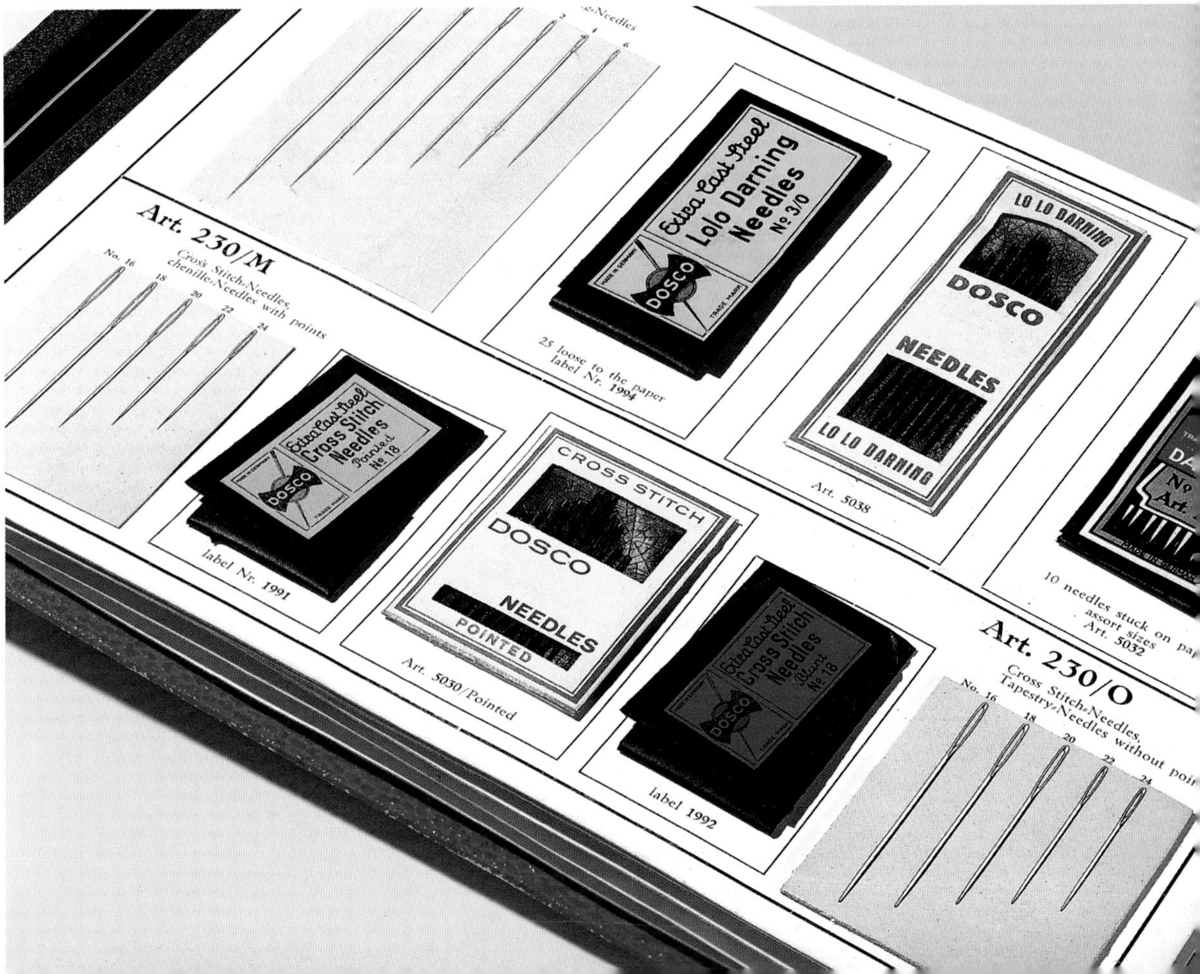

4

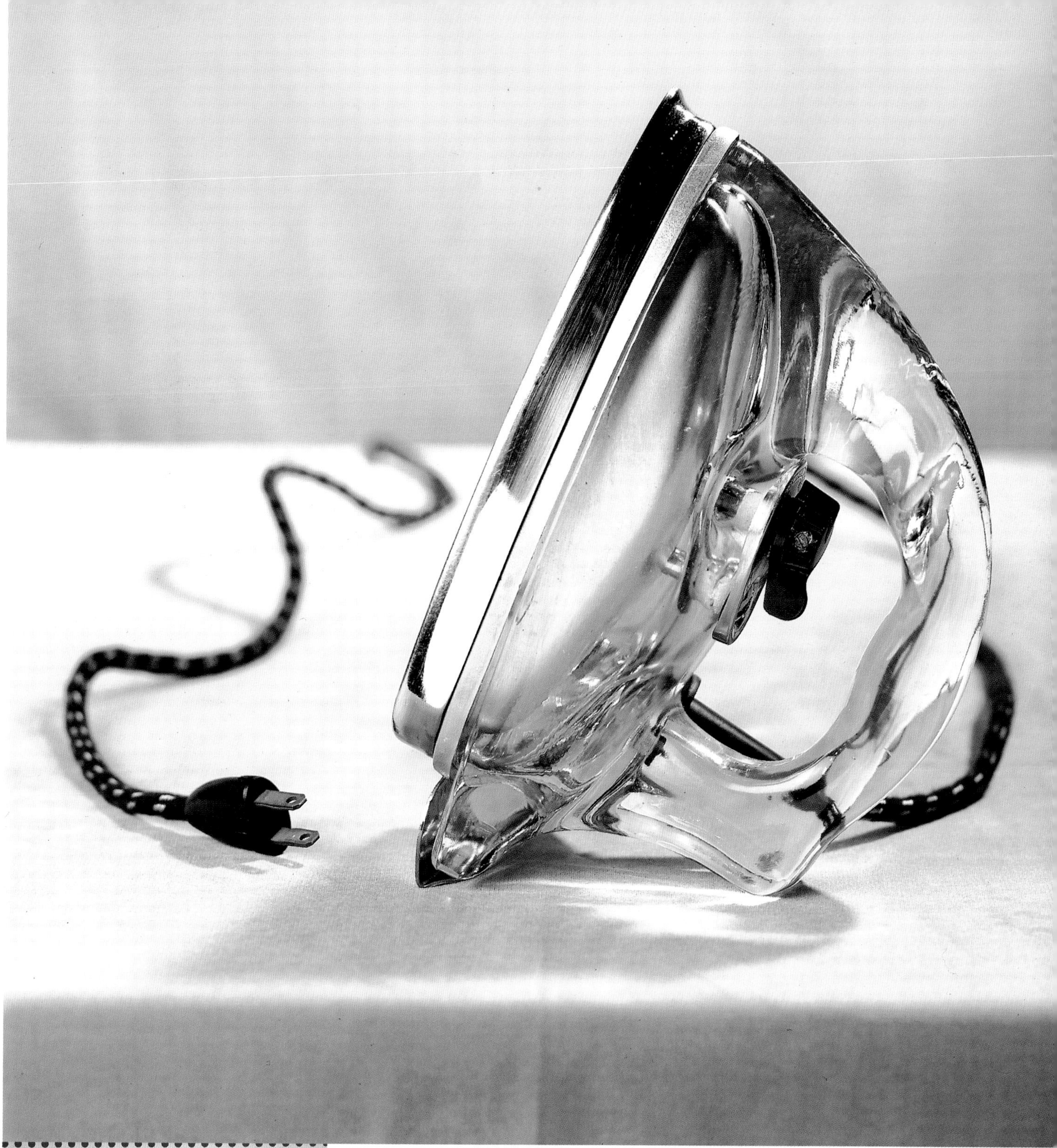

5

4 This glass iron was designed and produced during the early 1940s, when American metal supplies were being diverted to the war effort. It was only produced for a few years and came in colored as well as clear glass.

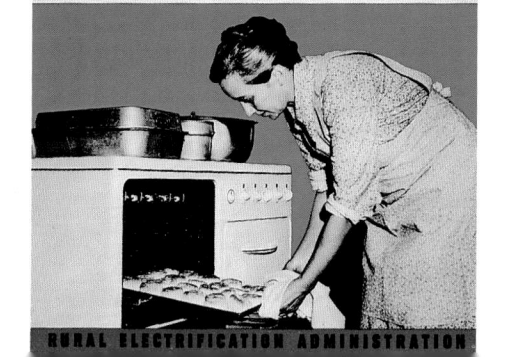

5 Lester Beall's graphically arresting poster done for the Rural Electrification Administration during the late 1930s suggests that the benefits of electricity can provide rural Americans with a wholesome, comfortable, and efficient lifestyle.

1

2

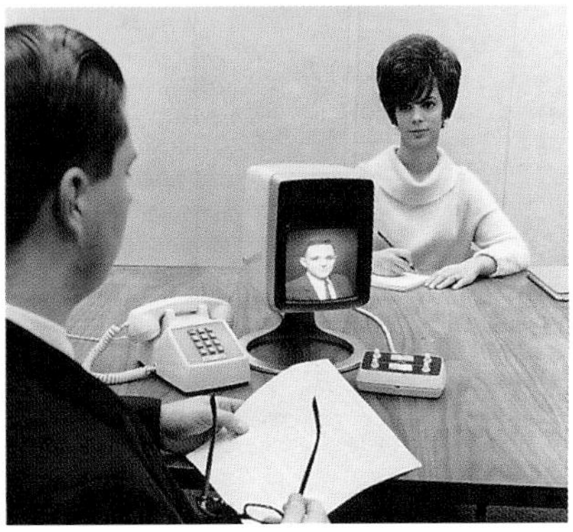
3

1 The Model #302 telephone designed by Henry Dreyfuss for Bell in 1937 remained the standard tabletop phone unit in America for decades. 2 *Hello—The Telephone at Your Service*, 1937, design for poster, Edward McKnight Kauffer for the British General Post Office. 3 Photograph of prototype videophone, Henry Dreyfuss Associates, 1967. 4 The 1919 Hammond Multiplex typewriter featured an innovative typeface system that allowed several different languages to be typewritten.

4

Let your fingers do the walking, the talking, and the working.

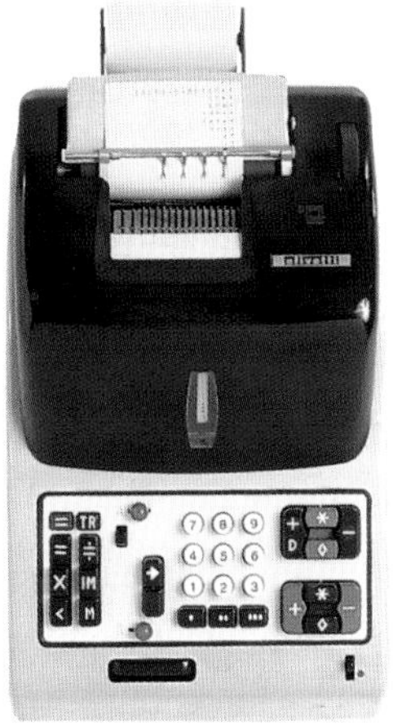

5

6

7

5 Marcello Nizzoli designed the Tetractys 24 electronic calculator for Olivetti in 1956. 6 The Valentine typewriter was designed by Ettore Sottsass and Perry King to be an "anti-machine machine," perfectly functional, yet meant to evoke a sense of fun. 7 frogdesign's NeXT computer, with its innovative Z-shaped base, contrasted with the prevailing white box-like forms of other computer units made in the mid-1980s.

1

1 Three imaginary tomb drawings, c. 1779-84, done by the visionary French artist Louis-Jean Desprez, known for his inventive theater designs. His use of Egyptian motifs marked a revival of European interest in Egyptian funerary decoration.
2 Léon Bakst's 1913 costume design for the queen's servant in Gabriel d'Annunzio's *Pisanella*.
3 Russian-born Serge Soudeikine's expressionistic squid costume design for the American premiere of Rimsky-Korsakov's production, *Sadko*, at New York's Metropolitan Opera in 1929.

2

3

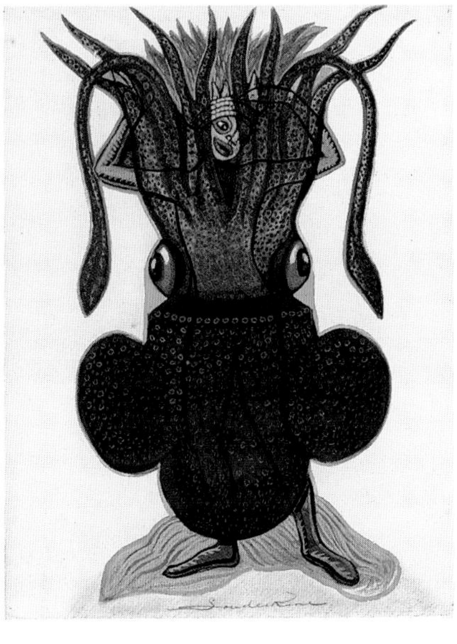

4

4 Alexandra Exter's 1921 drawing for a costume for a *boyar*, or nobleman, is a powerful example of Russian Constructivist design.

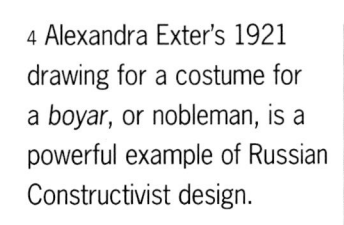

5 Three competition entries done in 1788 for La Fenice Theater, submitted by an unknown Italian architect. A historic landmark in Venice, La Fenice, originally designed by Gian Antonio Selva, survived until 1996, when fire destroyed this most celebrated of Italian theaters.

5

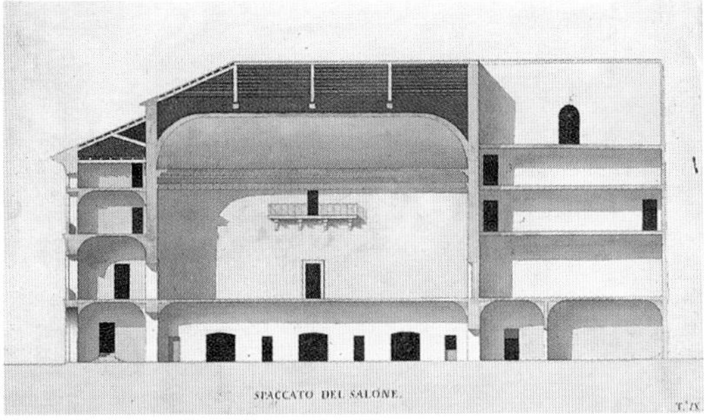

"The play's the thing..."

1

2

3

4

1 In 1964, Richard Sapper and Marco Zanuso designed the Ls502 portable radio for the Italian firm Brionvega, one of the first generation of portable "boomboxes." 2 "Jazz 3," a wallpaper designed by Cuno Fischer for the German market in 1959-60. 3 Milton Glaser and his colleagues at the Push Pin Studio in New York created a uniquely American graphic design style in the 1950s, '60s, and '70s that was imitated across the world. Glaser's work merges a bright, cartoon aesthetic with a rigorous sense of classical form. This 1978 poster uses overlapping fields of flat, opaque ink to construct an image of a jazz musician. 4 The big dials and expansive shape of the Crosley D25BE, an early clock radio, reflect the American fascination with the culture of the car in the 1950s.

Design dances to a different drummer with each generation.

5 Based in Seattle, Art Chantry has designed posters, magazines, and album covers for the city's vibrant music and art scenes since the late 1970s. Many of his designs, such as this poster from 1986, feature arresting montages and overlapping layers of transparent ink.
6 General Electric's SC 7300 stereo system was a limited-edition set produced around 1973 that included a record turntable, radio/tuner, eight-track tape player, and twin speakers, all housed in an elegant pedestal-form unit.
7 In the post-World War II period, companies like General Electric hired consultant industrial designers to work with their in-house design teams to create consumer products. In 1957 Richard Arbib incorporated the flaring side fins, flashy dial, and vivid colors of '50s automotive styling into the design of this clock radio.

5

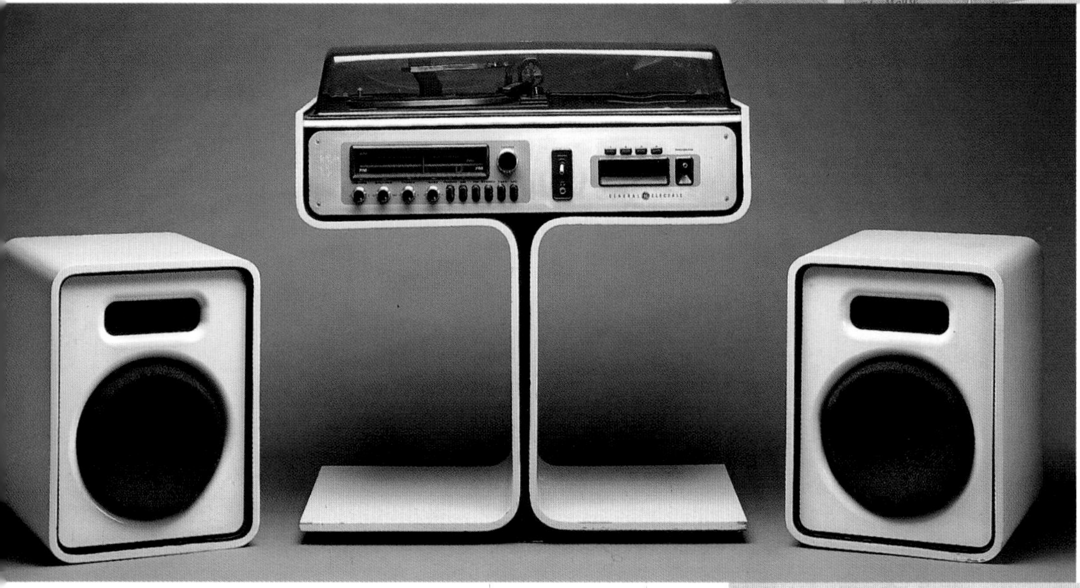

6

7

1

You ought to be in pictures.... Since the invention of photography in 1839, the proliferation of cameras as a mass market phenomenon has been unprecedented.

1 The Beau Brownie, with its modern geometric enameled decoration, was designed by Walter Dorwin Teague for the Kodak Company, c. 1930. 2 The Plus 126 camera was manufactured in the U.S.A. and Austria, c. 1970, of inexpensive, lightweight plastic and was designed to attach to a film cartridge. 3 The firm of Henry Dreyfuss worked on the design of the Polaroid Automatic 100 camera between 1961 and 1968. This drawing was made for the client company to show how the individual components were to be assembled to achieve a more compact camera body. Revolutionary in its inner workings and exterior housing, the camera weighed 50% less than the previous model and featured engineering changes that enabled the user to achieve a better image and extract the film more easily. It was Polaroid's first "pack film," versus roll-type film, camera.

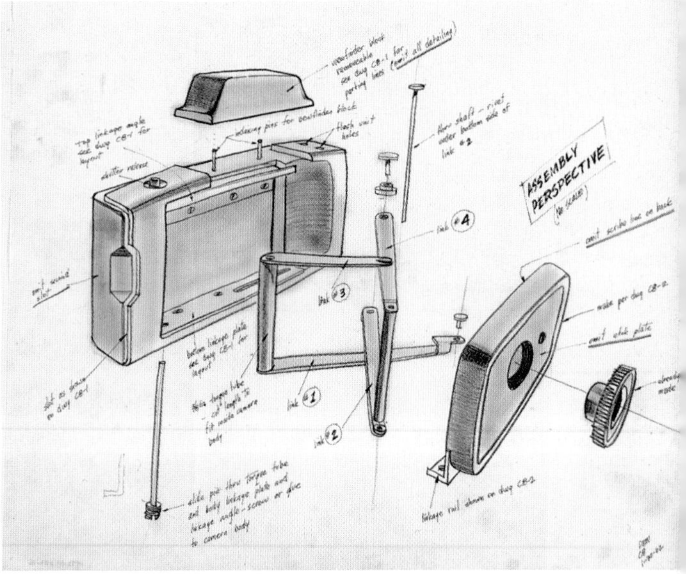

3

2

4

4 Raymond Loewy designed the Purma Special camera in 1937 for the English company Purma Cameras, Ltd. The first popular-market camera with an acrylic lens, it was significantly less expensive than cameras with optical glass. 5 A 1963 publicity photograph for the Polaroid Automatic 100 Land Camera.

LAND CAMERA
AUTOMATIC 100
POLAROID

1

With a simple surrender in paper and wood, we construct other worlds for ourselves and our children where we can leave the trappings of reality and abandon ourselves to play.

1 Commercial paper dolls such as this were introduced in America, c. 1850. More than a simple pastime, these dolls served to educate children about cultural and historical events, as well as to document fashion and taste.

2

2 Based on the Winnie the Pooh drawings by Ernest Shepard, this wallpaper frieze from the 1920s illustrates "Christopher Robin leads an Expotition (sic) to the North Pole."

3 This cut-and-paste paper model (c. 1870) for children features all the parts of a park kiosk. 4 A "cel" from the Beatles' 1968 film *Yellow Submarine*, animated by the German graphic designer Heinz Edelmann, mixes the influences of psychedelia, Pop art, surrealism, and Art Nouveau. 5 American puppeteer Tony Sarg designed the pattern of puppets and toys printed onto this children's fabric around 1935. 6 Hand-painted game boards, such as this French example, c. 1780, were popular forms of entertainment in 18th-century drawing rooms throughout Europe.

3

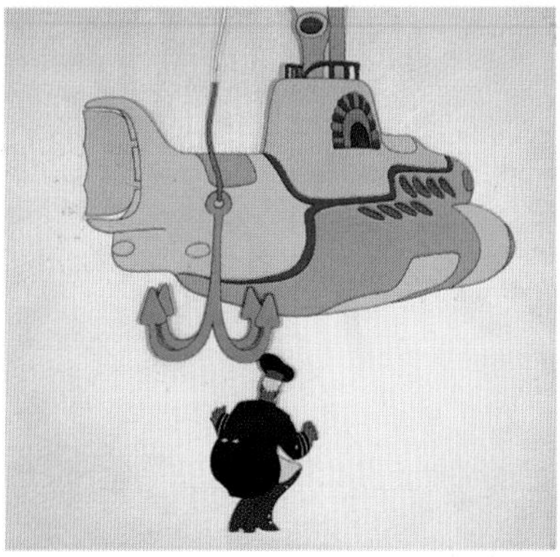

4

5

6

7 This 1933 *Pop-up Pinocchio...* is a delightful animated version of the popular 16th-century fairy tale, containing pop-up constructions by illustrator Harold Lentz.

Leisure beckons in images of play and pleasure.

1

2

3

1 An illustration of an iron settee manufactured in the late 19th century by the J. L. Mott Co. 2 After World War II, the German wallpaper industry resumed production in 1949. Popular culture was a rich source of motifs for wallpapers during this period, such as this beach scene pattern produced in 1958. 3 During the 1870s, Winslow Homer, one of America's greatest 19th-century painters, depicted women absorbed in daily life activities, such as this scene of a girl reading in a hammock. The play of light and shadow transforms this charming image into a painterly investigation of tonal contrasts.

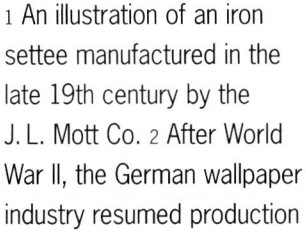

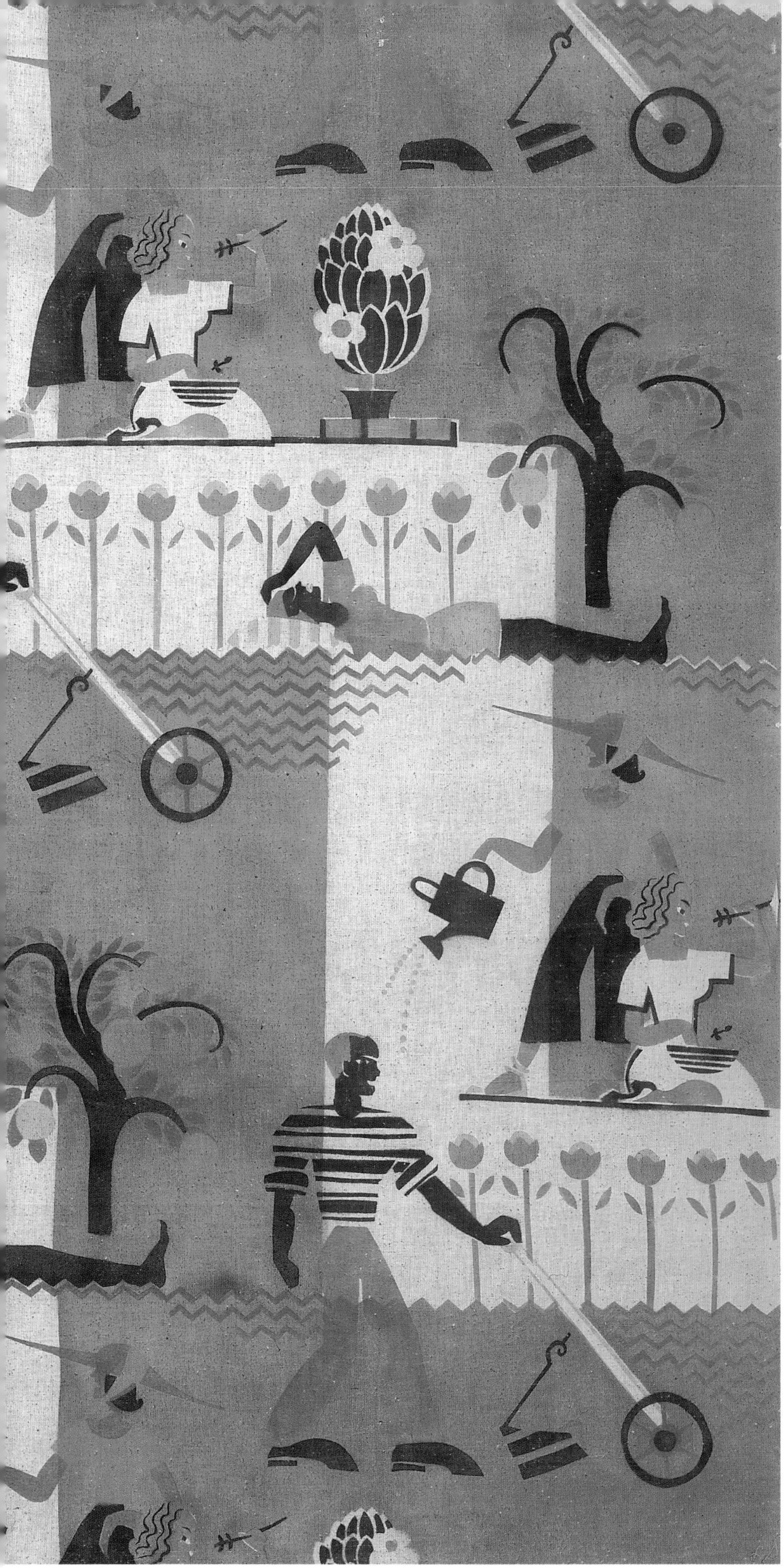

4

5

4 “Egyptian Garden” was designed and stenciled on fabric by American designer Lanette Scheeline, c.1939. The everyday tasks of gardening are treated in the style of an ancient Egyptian wall painting. 5 The woman swimmer featured at the center of this plate celebrates the sports and fitness campaigns sponsored by the Soviet government during the 1930s. The border surrounding the plate includes ornamental “USSR” monograms and vignettes showing gymnasts, runners, and other female athletes.

1

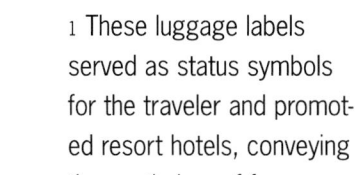

1 These luggage labels served as status symbols for the traveler and promoted resort hotels, conveying the exotic lure of faraway places. 2 The advertising message of Edward McKnight Kauffer's 1948 poster is clear and direct: American Airlines rules the sky. The logo appears like a patriotic sign on a soaring New York City skyscraper. 3 This 1972 poster for the International Design Conference in Aspen vividly illustrates the transportation network that links the world's cities into a vast megalopolis. Ivan Chermayeff, a leader in the field of corporate identity as well as the creator of numerous posters for cultural events, is known for his distinctive use of collage and found objects.

2

Come fly with me!

3

Airports are a link to the people and places of the world.

Airports are at a crossroads where all nations are represented by their own airlines, carrying passengers of all nationalities, races and religions.

The 25 busiest airports in the United States serve over 200,000,000 passengers a year, carrying them to and from many thousands of destinations in every continent. People traveling for pleasure, for business, and sometimes to relocate permanently in new countries around the globe.

Airports are places to see the newest methods and systems for moving material and people safely and quickly.

People speak many languages, and so the airlines and airports have adopted, by international agreement, common codes, signs and symbols of their own to ease the flow of cargo, people and baggage.

Airports give one the opportunity to learn, not only about aircraft and air travel, but about how the people of the world communicate with one another.

The following gives the city, its country and currency of the listed airport codes:

ABJ: Abidjan, Ivory Coast, CFA Franc
ACC: Accra, Ghana, New Cedi
AMS: Amsterdam, Netherlands, Guilder
ANK: Ankara, Turkey, Lira
ANU: Antigua, British West Indies, Pound
BCN: Barcelona, Spain, Peseta
BDA: Bermuda, Bermuda, Pound
BDI: Barbados, British West Indies, W. I. Dollar
BEG: Belgrade, Yugoslavia, Dinar
BEL: Belem, Brazil, Cruzeiro
BER: Berlin, Germany, Deutsche Mark
BEY: Beirut, Lebanon, Lebanese Pound
BKK: Bangkok, Thailand, Baht
BRU: Brussels, Belgium, Franc
BSB: Brasilia, Brazil, Cruzeiro
BUE: Buenos Aires, Argentina, Peso
CMN: Casablanca, Morocco, Dirham
COO: Cotonou, Dahomey, CFA Franc
CPH: Copenhagen, Denmark, Krone
DAM: Damascus, Syria, Syrian Pound
DAR: Dar es Salaam, Tanzania, Tanzanian Shilling
DIA: Dulles Int. Airport, Wash., D.C., U.S., Dollar
DKR: Dakar, Senegal, CFA Franc
DLA: Douala, Cameroons, CFA Franc
DUS: Dusseldorf, Germany, Deutsche Mark
EBB: Entebbe, Uganda, East African Shilling
FAI: Fairbanks, Alaska, United States, Dollar
FDF: Fort-de-France, Martinique, Franc
FIH: Kinshasa, Republic of Congo, Congo Franc
FPO: Freeport, Bahamas, Dollar
FRA: Frankfurt, Germany, Deutsche Mark
GEO: Georgetown, Guyana, Guyana Dollar
GLA: Glasgow, Scotland, Pound
GUA: Guatemala City, Guatemala, Quetzal
GUM: Guam, Guam, Dollar
HAM: Hamburg, Germany, Deutsche Mark
HEL: Helsinki, Finland, Mark
HKG: Hong Kong, China, Yuan
IST: Istanbul, Turkey, Lira
JNB: Johannesburg, Union of South Africa, Rands
KEF: Keflavik, Iceland, Krona
KIN: Kingston, Jamaica, Pound
KHI: Karachi, Pakistan, Pakistan Rupee
LIS: Lisbon, Portugal, Escudo
LON: London, England, Pound
LOS: Lagos, Nigeria, Pound
MBJ: Montego Bay, Jamaica, Pound
MIA: Miami, Florida, United States, Dollar
MIO: Maracaibo, Venezuela, Bolivar
MOW: Moscow, Soviet Union, Ruble
MUC: Munich, Germany, Deutsche Mark
MVD: Montevideo, Uruguay, Peso
NAS: Nassau, Bahamas, Dollar
NBO: Nairobi, Kenya, East African Shilling
NCE: Nice, France, Franc
NDH: New Delhi, India, Rupee
OSA: Osaka, Japan, Yen
OSL: Oslo, Norway, Krone
PAR: Paris, France, Franc
PBM: Paramaribo, Venezuela, Bolivar
POS: Port of Spain, Trinidad, Trin. & Tob. Dollar
PRG: Prague, Czechoslovakia, Koruna
PTP: Pointe-a-Pitre, Guadeloupe, Franc
RGN: Rangoon, Burma, Kyat
RIO: Rio de Janeiro, Brazil, Cruzeiro
ROM: Rome, Italy, Lira
RSD: Rock Sound, Bahamas, Dollar
SAL: San Salvadore, El Salvador, Colon
SAO: Sao Paolo, Brazil, Cruzeiro
SDQ: Santo Domingo, Dominican Republic, Peso
SJU: San Juan, Puerto Rico, Dollar
SMA: Santa Maria, Azores, Escudo
SNN: Shannon, Ireland, Irish Pound
STO: Stockholm, Sweden, Krona
STR: Stuttgart, Germany, Deutsche Mark
STT: St. Thomas, Virgin Islands, Dollar
STX: St. Croix, Virgin Islands, Dollar
THR: Teheran, Iran, Rial
TYO: Tokyo, Japan, Yen
VIE: Vienna, Austria, Schilling
WAW: Warsaw, Poland, Zloty

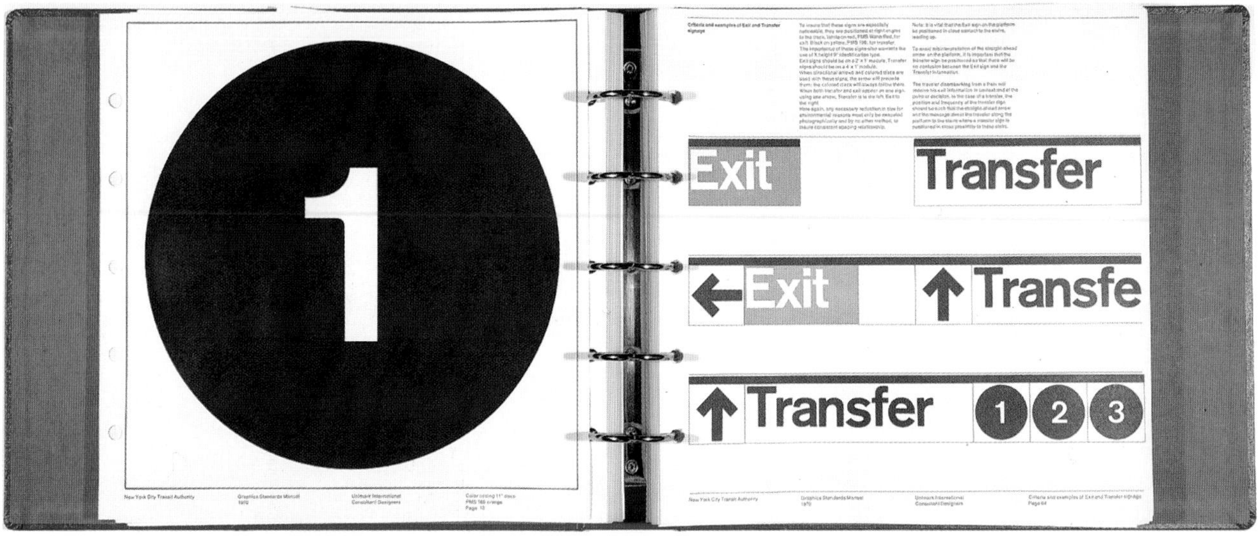

1

1 Massimo and Lella Vignelli redesigned the graphics and signage for the New York City Transit Authority subway system and, in 1970, produced a manual for the proper use of color, spacing, and lettering to conform to their design. 2 Dining car appointments for the 20th Century Limited train, Henry Dreyfuss, 1938. 3 Henry Dreyfuss Associates relied on extensive ergonomic research in designing a seat for Lockheed Corporation's Electra 188 airplane in 1955. The x-ray photograph shows a figure seated in a mock-up of a seat, and the schematic drawing shows the seat back with the head area formed without a contoured pillow. 4 Comfort, color, and convenience were priorities in the design of the Convair 880 airplane lavatory designed by Dorothy Draper & Co. in 1957.

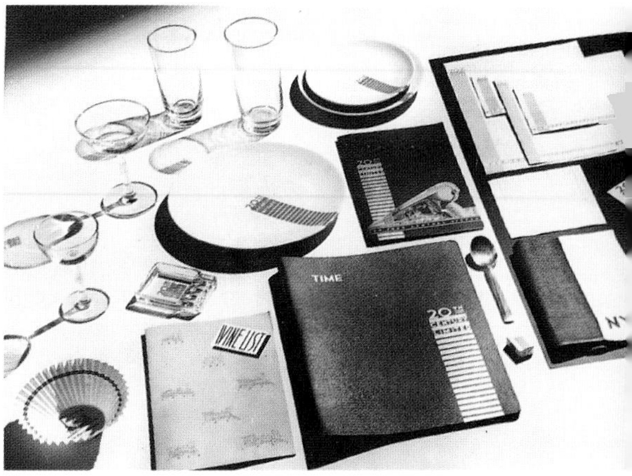

2

Creature comforts for 20th-century nomads.

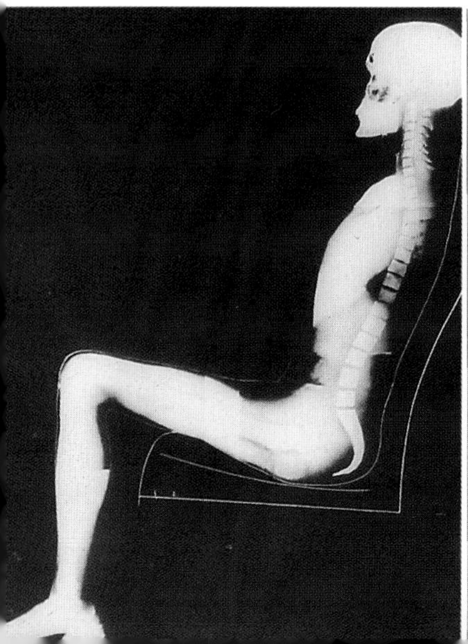

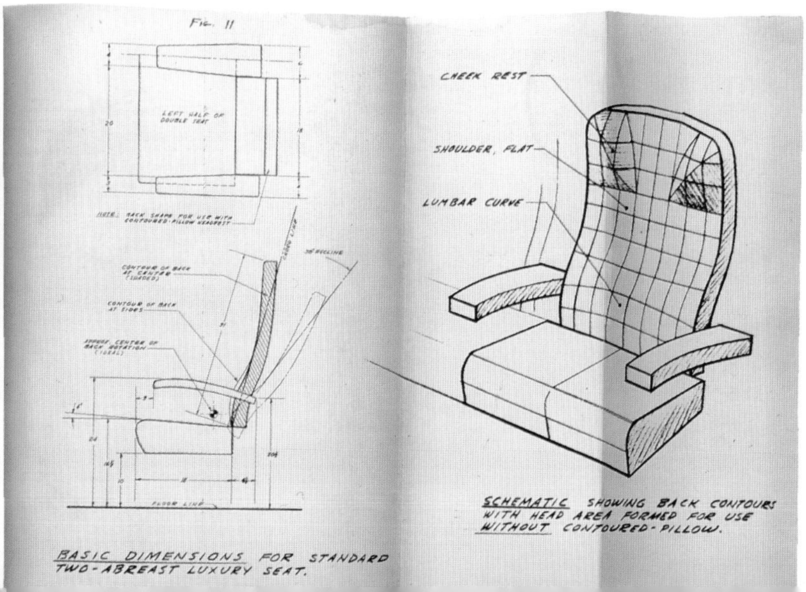

3

4

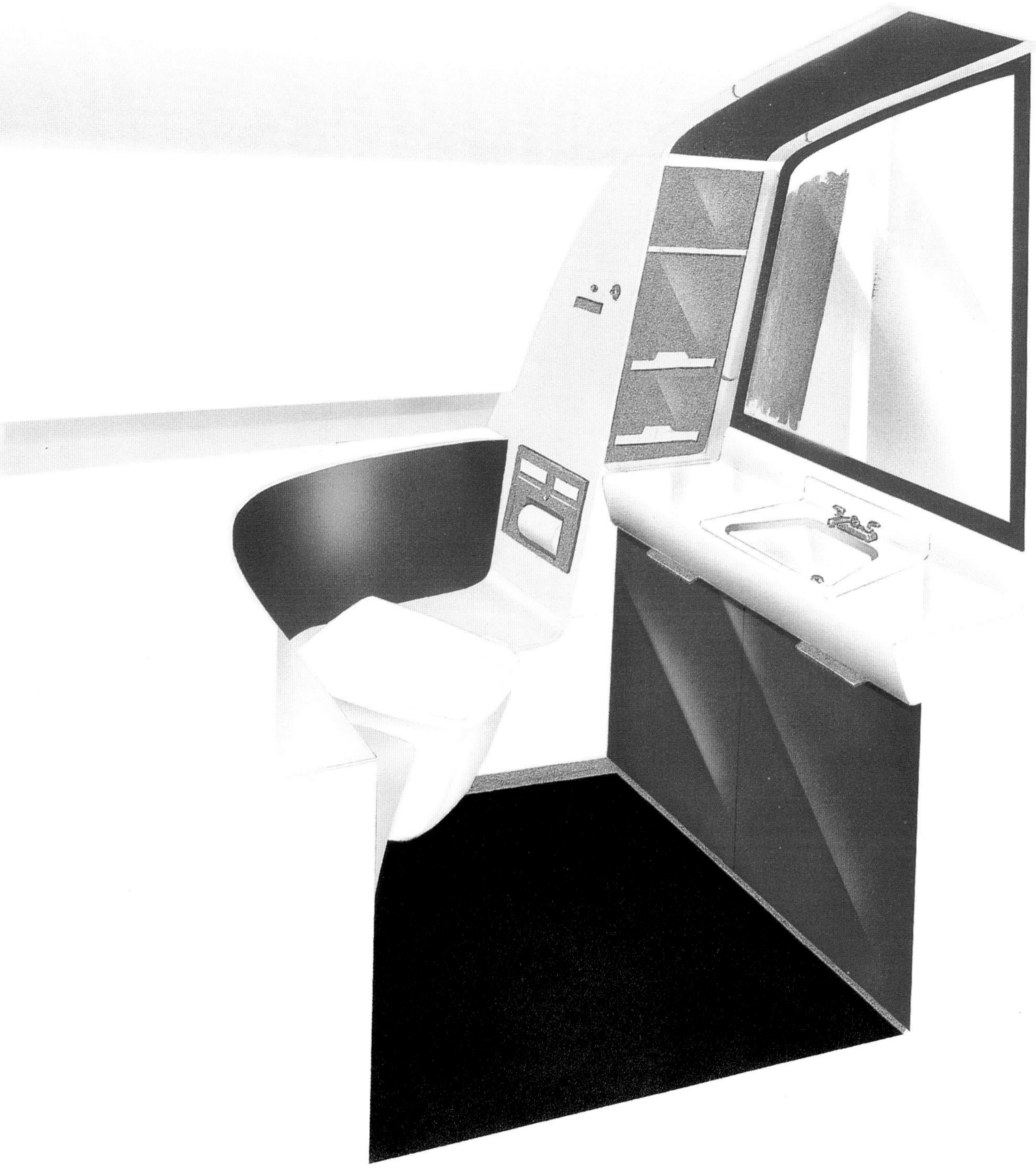

Styled for speed

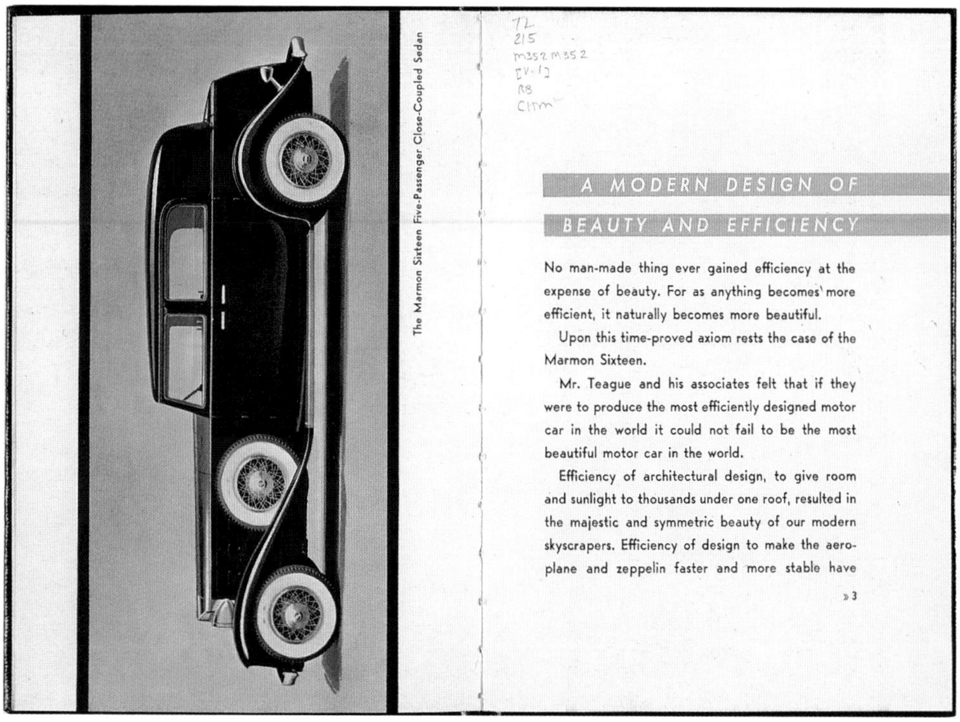
The Marmon Sixteen Five-Passenger Close-Coupled Sedan
A MODERN DESIGN OF
BEAUTY AND EFFICIENCY
No man-made thing ever gained efficiency at the expense of beauty. For as anything becomes more efficient, it naturally becomes more beautiful.
Upon this time-proved axiom rests the case of the Marmon Sixteen.
Mr. Teague and his associates felt that if they were to produce the most efficiently designed motor car in the world it could not fail to be the most beautiful motor car in the world.
Efficiency of architectural design, to give room and sunlight to thousands under one roof, resulted in the majestic and symmetric beauty of our modern skyscrapers. Efficiency of design to make the aeroplane and zeppelin faster and more stable have
» 3
1

1 In 1930, Walter Dorwin Teague published this promotional book on his firm's design of the Marmon Sixteen sedan, notable for its form as well as its use of the light and durable wonder metal, aluminum. 2 The Convair Autoplane, designed by Henry Dreyfuss in 1947, was an experimental vehicle for uninterrupted travel between air and land. Upon landing, the auto disengaged from the plane and the pilot could drive to the final destination. 3 The cover of *Maps...and how to understand them*, prepared by Richard E. Harrison, J. Mca. Smiley, and Henry Lent in 1943, showing air travel routes from the United States to foreign cities.

2

MAPS
. . . and how to understand them
CONSOLIDATED VULTEE AIRCRAFT CORPORATI
3

4

5

4 This clay scale model of the 1936 S-1 Locomotive designed by Raymond Loewy was tested more than 100 times to observe the airflow around the engine. Adjustments to the roof smoke deflector and other streamlining features reduced the air resistance by 35% over other engines, enabling the locomotive to move at optimum speed.

5 A 1936 drawing for the S-1 Locomotive. This 6,000 horsepower locomotive was used to pull the Broadway Limited, the most famous train Loewy designed for the Pennsylvania Railroad.

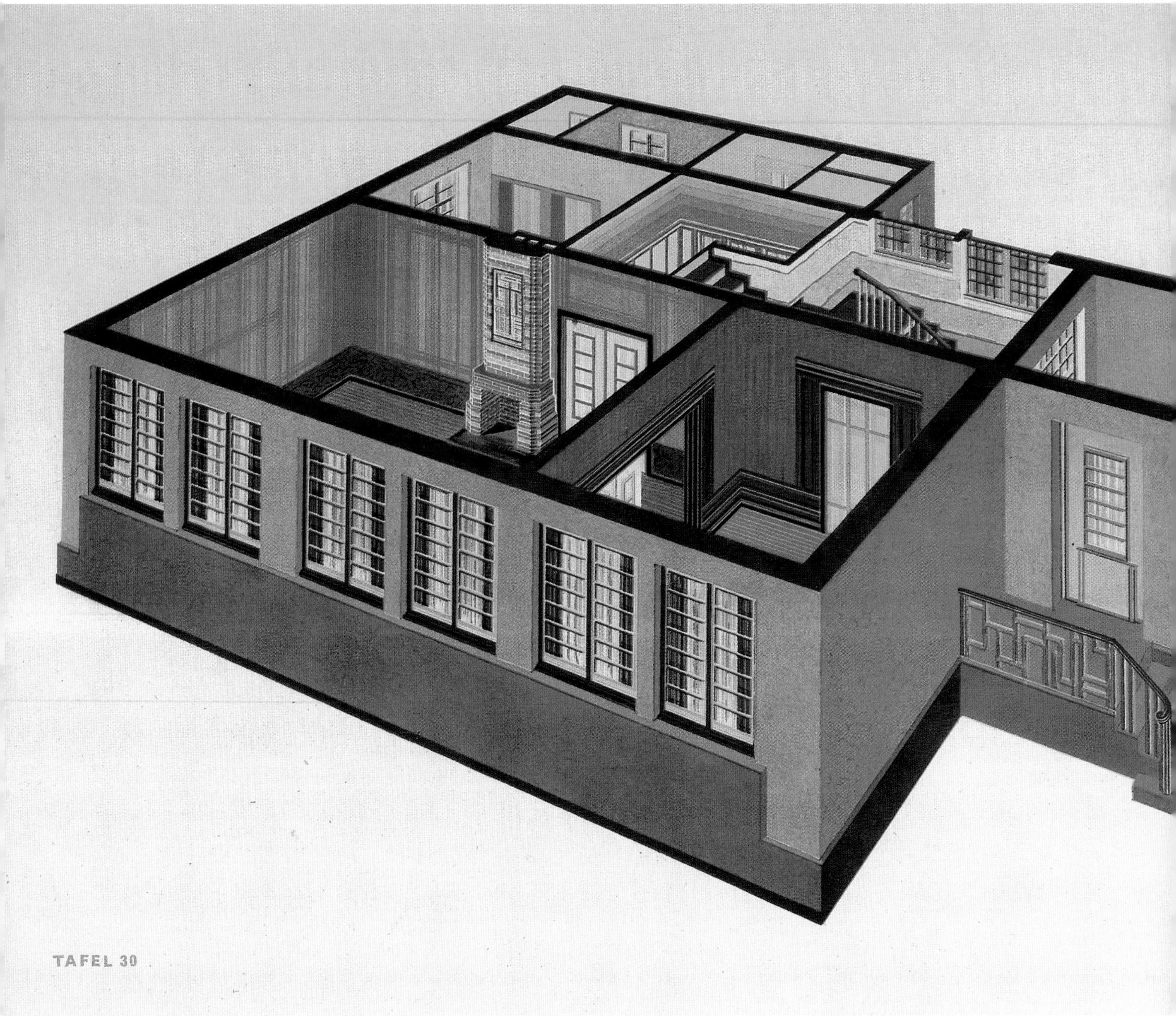

A plate from *Farbige Räume und Bauten* (Color in Rooms and Buildings) by Wilhelm Jöker, c. 1929, published as a guide for painters, architects, and designers to demonstrate possibilities for the manipulation of space through the controlled and extensive use of color.

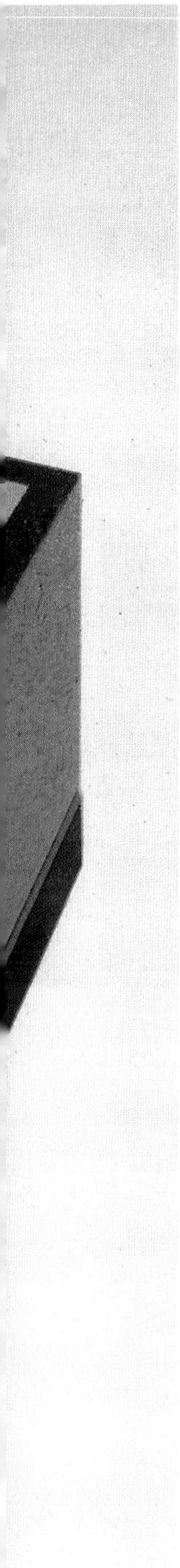

Design for Shaping Space

The work of shaping space is full of implications about our relations to each other and to nature at any given time. The degree to which we value community, privacy, security, sociability, and social status is evident in the design of building types—from palaces to prisons—and in their spatial configurations—from cities to suburbs. In shaping space, we also affect the environment. Just as societal patterns have shaped our dwellings, our attitudes toward nature have conditioned the way we approach the landscape. Parks, zoos, gardens, preserves—even roads, fences, benches, and fountains—historically reflect our changing views of the natural.

R. Buckminster Fuller saw new possibilities for housing in high rise technology. A design for a "Ten-Deck House" by Fuller shows an aluminum tower supported by cables, conceived as a prototype for a single-family, low-cost housing unit. The tower was to be factory-assembled and transported by blimp to its foundation crater. Although he was a zealous advocate of scientific solutions to social and environmental issues, even the economy-minded Fuller was not immune to the romance of building upward. In the Museum's drawing, the "Ten-Deck House" features a sky promenade at its pinnacle and is depicted as though it were a rocket poised for take-off.

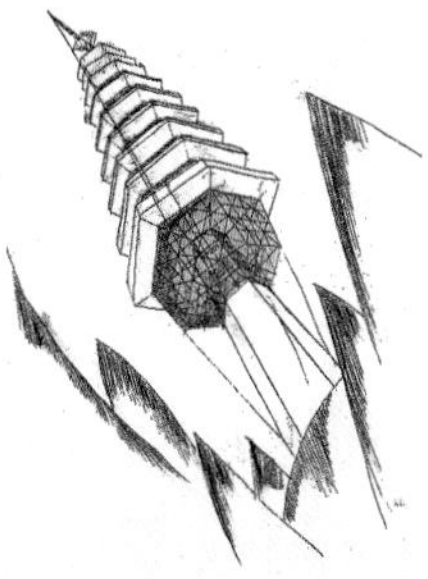

"Ten-Deck House" by R. Buckminster Fuller, c. 1928.

While Fuller had the working family in mind, Donald Deskey was thinking about their vacations when he envisioned the "Sportshack" in 1940. Deskey, best known as the interior designer of Radio City Music Hall, was also a proponent of prefabrication. During his career as an industrial designer he created schemes for modular homes that reflected his life-long fascination with new materials. His speculative drawing for the "Sportshack" features walls of glass that would have allowed ample views of nature: the vernacular of the American get-away cabin meets the modernism of the European glass pavilion.

"Sportshack" by Donald Deskey, 1940.

Twentieth-century modernism championed the idea of bringing the outdoors inside with large, unbroken expanses of window providing generous vistas. Irregular nature was the ideal foil for the rational, rectilinear presence of architecture. A little over one hundred years earlier, the English landscape designer Humphry Repton also considered the landscape in relationship to the architecture of his day. However, instead of contrast, he sought harmony between a setting's structures and its plantings.

A skilled artist, Repton created handsome leather-bound books of his schemes, such as the 1803 volume *Observations on the Theory and Practice of Landscape Gardening*. These ingenious volumes, embellished with hand-colored plates, featured overslips to show "before" and "after" views to persuade the prospective client. Repton supplied the English squirearchy—a rising, affluent middle class—with a manicured version of nature that he felt befitted their social standing. His illustrations pictured long, expansive lawns stretching to gently sloping horizons of trees that masked plowed fields and other evidence of labor. Views of nature were necessary for the contemplative life of the country gentleman. Direct experience, however, was to come from the fragrance of flower gardens planted close to the house—one of Repton's chief innovations—and was mediated by an architecture of terraces, promenades, and conservatories.

Humphry Repton's 1803 landscape gardening folio.

Just as outdoor "rooms" such as gardens, greenhouses, and zoos are cultural creations, indoor rooms also reflect social notions of space. In traditional Japanese homes, screens and draped fabric create movable walls, permitting several different activities to take place in the same room in the course of a day. Time and space are considered inextricable concepts and are joined in the Japanese word *ma*.

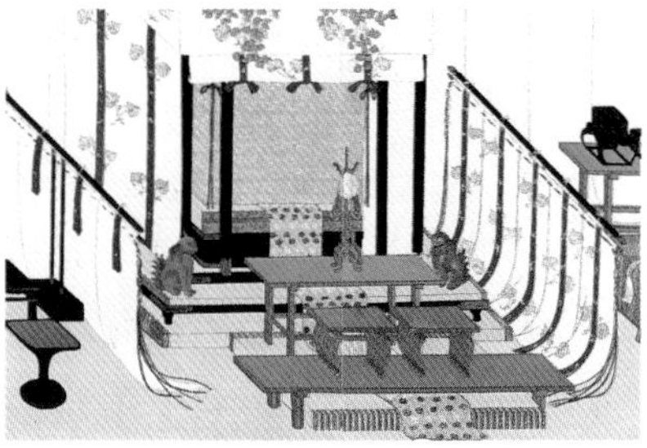

Illustration from the 1903 book *Sixteen Japanese Ceremonies*.

During the 18th century, traditional western homes evolved a hierarchical use of the interior, with functions of eating, sleeping, and entertaining segregated into halls, parlors, bathrooms, bedrooms, kitchens, and dining rooms. While the configurations in which they are ordered are continually changing—today, cooking and eating, and entertaining and working, are often collapsed into common spaces—the wall has retained its preeminence in defining where we are when we are at home or at the office. So central is it to our understanding of architecture, our notions of privacy and community, that we have devised endless treatments of its surface. Walls are like the pages of a book on which we inscribe our identities and tell our stories.

Panoramic wallpapers are among the most dramatic and complex expressions of our appetite for visual entertainment, presaging the large-screen televisions of today's media rooms with their enormous scale and sequential panels. The 19th-century masterpiece "Psyche" is a stunning tour de force in the genre of wallpaper panoramas. Each tableau features a scene from the ancient Greek myth in which the jealous Venus enlists her son Cupid to banish the beautiful Psyche.

A scene from "Psyche" wallpaper panorama, 1815-16.

(The scheme backfires, and, as events unfold, Cupid and Psyche fall in love. They are thwarted by Venus and their own foibles, but finally are reunited.) The story was highly popular in the 19th century, and the recent invention of scenic wallpapers in France made it possible to depict it episodically with life-size imagery. Rendered in *grisaille*—shades of black, gray, and white—by Merry-Joseph Blondel and Louis Lafitte, the fable of Cupid and Psyche is told in 320 square feet of wallpaper. Comprised of twenty-six panels, it was first printed in 1816 from an astounding 1,245 separate wood blocks by Joseph Dufour, a leading producer of French wallpaper.

Detail of "Acanthus" wallpaper by William Morris, 1875.

As spectacular as narrative papers such as "Psyche" are, they are relatively rare theatrical transformations of space. More familiar are the patterned papers that complement the furnishing schemes of their rooms. Victorians favored the conceit of imitation, and the late 19th and early 20th centuries witnessed a host of stylistic revivals. Gilded, embossed, and stenciled imitations of early Dutch leather wallcoverings graced neo-Renaissance salons and libraries. *Faux bois* papers masked plaster walls with the warmth and pattern of simulated wood grains Pictorial imagery, however, never lost its appeal. Large ferns and florals were especially popular in the 1930s and '40s. The postwar 1950s saw a burst in the production of wallpapers with repeat patterns featuring scenes of Paris and Rome, now familiar to ex-GI's. The preoccupations of the decade were also reflected in a lively assortment of popular themes, from hot rods to space travel.

Wallpaper with images of space stations and rockets, c. 1950.

If walls are the tablets that record our tastes, they are also the containers of our possessions—the chairs, beds, tables, desks, lamps that form the armature of our days. The bed, in particular, has at times approximated a small room, in and of itself. The site of romantic encounters, a haven of respite at day's end, and, in the past, a platform for royal rituals, the bed at various times has grown canopies and curtains to envelop its occupants in requisite privacy or surround them in regal finery.

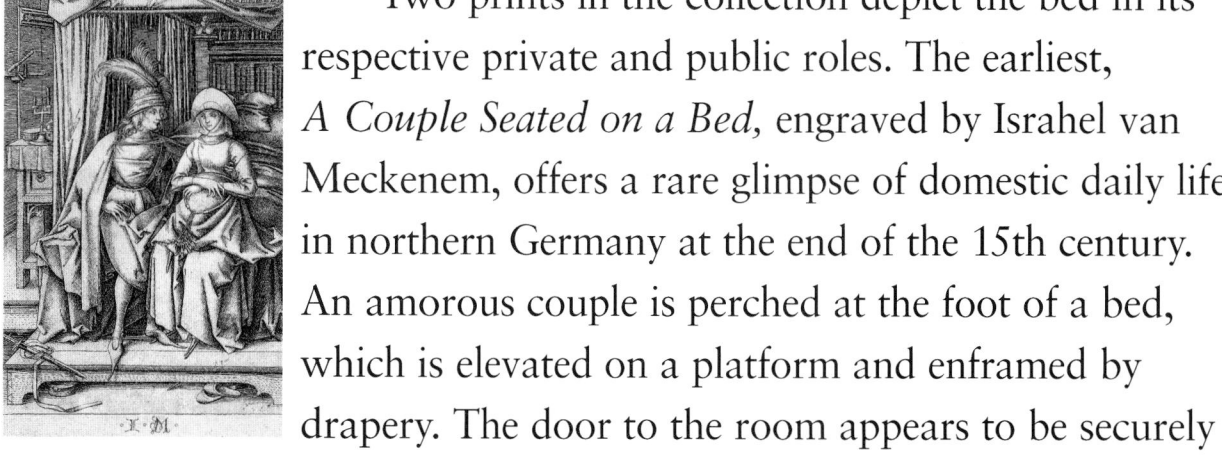

A Couple Seated on a Bed by Israhel van Meckenem, c. 1500.

Two prints in the collection depict the bed in its respective private and public roles. The earliest, *A Couple Seated on a Bed,* engraved by Israhel van Meckenem, offers a rare glimpse of domestic daily life in northern Germany at the end of the 15th century. An amorous couple is perched at the foot of a bed, which is elevated on a platform and enframed by drapery. The door to the room appears to be securely latched to permit no untoward interruption of the intimacies implied.

In contrast to the bourgeois marriage bed, the state beds of kings and queens were the most formal and public of places. Though used only on rare occasions, they were permanently on display as a visible manifestation of royal wealth and status. A print by Daniel Marot, the exiled French Huguenot designer associated with the William and Mary style of 18th-century England and Holland, gives a detailed look at the centerpiece of the state bedchamber. The hallmarks of the bed are its great canopy decorated with ostrich plumes at the corners, a massive carved headboard, and sumptuous upholstery. The ensemble would have stood over fifteen feet high, and while its drapery could be closed, a bed such as this would have doubtless served more affairs of state than affairs of the heart.

Design for a state bed-chamber by Daniel Marot, c. 1702.

If warmth were brought into the regal sleeping chamber, it was more likely to come from the embers of the fireplace or the glow of a sconce. A cold volume of space becomes a habitable room with an

infusion of heat and light. Designers have fashioned an astounding array of artifacts—thermostats, andirons, hearths, candelabra, chandeliers, desk lamps, flashlights—to contain and disperse these intangible necessities. Because of the very immateriality of the energy they control, the shapes of these objects often contain symbols or metaphors for the purposes they serve. A pair of gilt bronze firedogs, or andirons, from the French court of Louis XVI features decorative motifs of flaming urns and crossed torches in the neo-classical style that prevailed in the 1780s. A century later, the harnessing of electricity radically transformed the field of lighting, and in 1957, Danish designer Poul Henningsen reinvented the chandelier with his delicately splayed "Artichoke" lamp. Its overlapping, copper-colored "leaves" spread warm light at the same time that they eliminate any possible glare, in a sophisticated fusion of form and function.

One of a pair of andirons attributed to Pierre Gouthière, c. 1775.

"Artichoke" lamp by Poul Henningsen, 1958.

Perhaps nowhere do the two closely linked concepts of form and function come into play more directly than in the design of a chair. The American designers Charles and Ray Eames played with the balance of those twinned principles throughout their practice, which spanned from 1941 until 1978. Despite the fact that the Eameses were prodigious designers of films, exhibitions, toys, and environments, they are still best known as designers of chairs. Their search for new forms for mass-produced, affordable seating yielded the famous molded fiberglass and plywood chairs that are the signature works

Rocking chair by Charles and Ray Eames, c. 1950.

of their career. Although laminated plywood had been in use since the mid-19th century, it was the Eameses who invented the technology for a plywood that could be bent in more than one direction to achieve complex curves. Ironically, one of the most light-hearted pieces in their series of plywood chairs—the 1944 child's chair—was made with the same tooling used for the wooden splints and aircraft parts their studio had invented during World War II. Pragmatism and sentiment meet in the cut-out heart on the back of the child's chair that serves as both handle and ornament.

Child's chair by Charles and Ray Eames, 1944.

Furniture—by nature of its volume, scale, and expanse of surface—has been a natural site for decoration, whether applied or integral to the form. Tables, desks, sideboards, and bureaus have invited carpenters, goldsmiths, and stonemasons to exercise their genius in details from elaborate drawer pulls to secret compartments. One of the most unusual instances of decorative invention can be found in the porcelain cladding of the Museum's stately jewel cabinet produced by the royal factory at Sèvres in 1826. Intricate depictions of gods and goddesses, garlands, roses, birds, and jewelry suggest the preciousness of the contents it might have held and the status of the royal personage who commissioned it. A diplomatic gift of state from Charles X of France to Francis I, King of the Two Sicilies, the elaborate cabinet and its porcelain plaques also served as a subtle advertisement of the superiority of French luxury goods.

Sèvres jewel cabinet, 1824-26.

In contrast to the distinctively articulated mass of the Sèvres cabinet is a design for a delicate mirror frame bearing the monogram of Marie Antoinette that is represented in the Museum's collection in a drawing by Richard de Lalonde from the late 1780s. The infinite space of reflection is crowned by naturalistic flowers and the initials of the famed queen for whose apartments it was designed. In its simplicity, the frame demurs from competing with the visage it might have held.

Design for a royal mirror by Richard de Lalonde, late 1780s.

Mirrors, windows, doorways, hallways, even screens and monitors all frame our views, teasing us to change them, to make them new again and, in the process, renew ourselves. It is when we enter the picture ourselves and sense the presence of our possessions or, sometimes, the void of their absence, that we begin the task of shaping space.

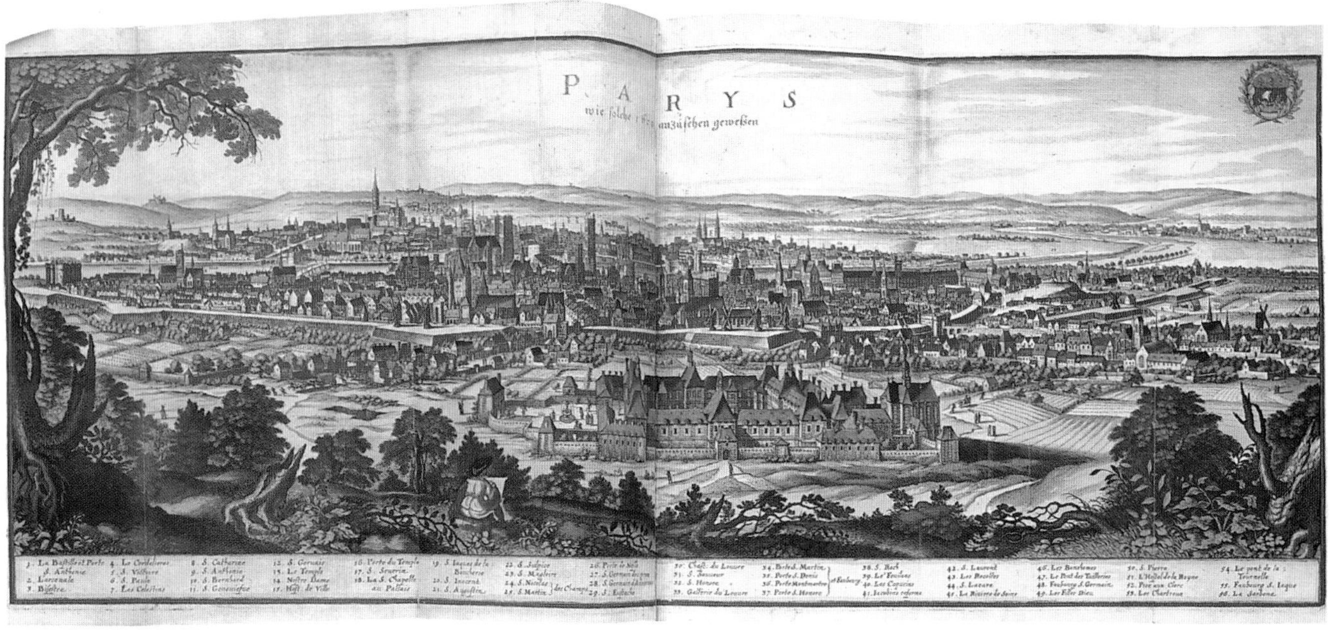

1

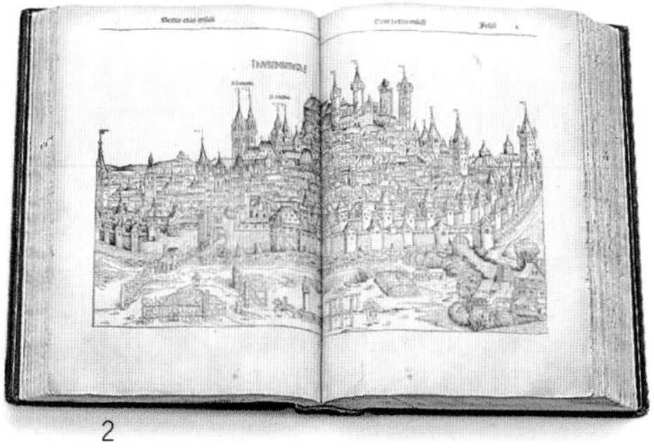

2

1 View of Paris from Martin Zeiller's *Topographia Galliae...*, 1655. 2 A woodcut image of the cityscape of Nuremberg from the famous Nuremberg Chronicle, 1493. 3 A plate from a mid-17th-century book by Johannes Nieuhof showing the plan of Canton, China. Nieuhof was a member of a Dutch trade mission dispatched to China after the fall of the Ming dynasty in 1644.

4 A zoning map of Manhattan from the 1969 Plan for the City of New York that sought to shape the city through new land use regulations. 5 Housed in the Perisphere of the 1939 New York World's Fair, the *Democracity* exhibition was designed by Henry Dreyfuss to introduce fair-goers to the city of the future.

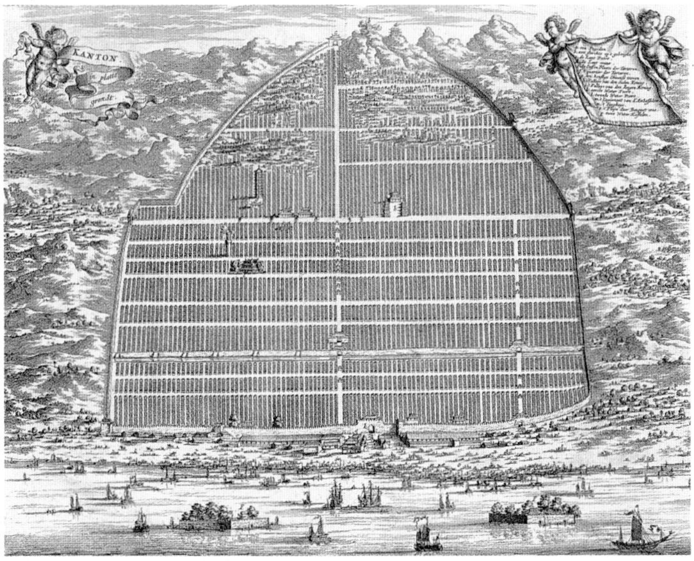

3

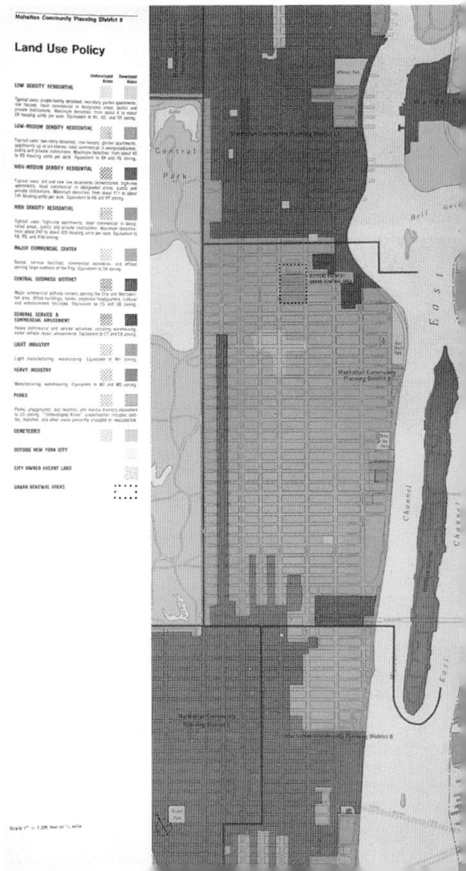

4

The scheme of a city gives evidence to the ideas of social order. Cities, once circumscribed by walls, radiated outward to suburban satellites with the advent of the automobile in the early 20th century.

5

3

4

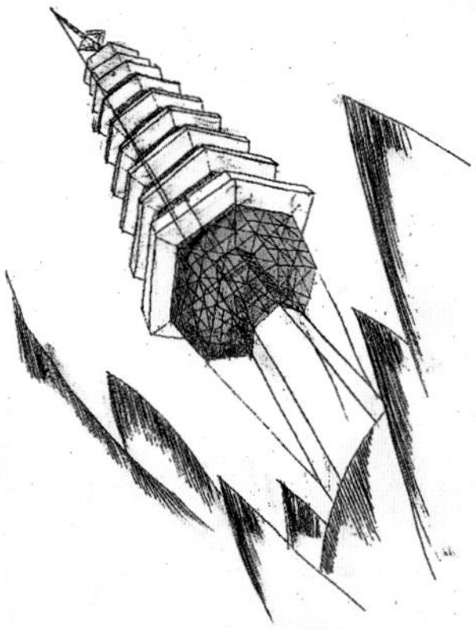

2

1 In 1922, Hugh Ferriss, one of America's foremost architectural renderers, produced this dynamic drawing. The design is a response to the limitations imposed by the landmark 1916 New York zoning law, which required that buildings be set back to admit more light to the city's streets.

2 The height of the Singer Building, center, compared to the heights of world landmarks, 1908. 3 Architect Paul Rudolph's 1980 design for the Colonnade Condominiums in Singapore juxtaposes suspended rooms with recessed spaces. Conceived by the architect as an animated assemblage of geometric volumes, the skyscraper reverberates with light and shadow.

4 R. Buckminster Fuller's design for the "Ten-Deck House," c. 1928, depicts a multi-level, tower-like building supported by cables. The futuristic building looks like a rocket poised for take-off. Fuller intended his "Ten-Deck House" to be factory-assembled to eliminate the cost of on-site construction; however, it was never realized.

Reaching for the Sky

1

1

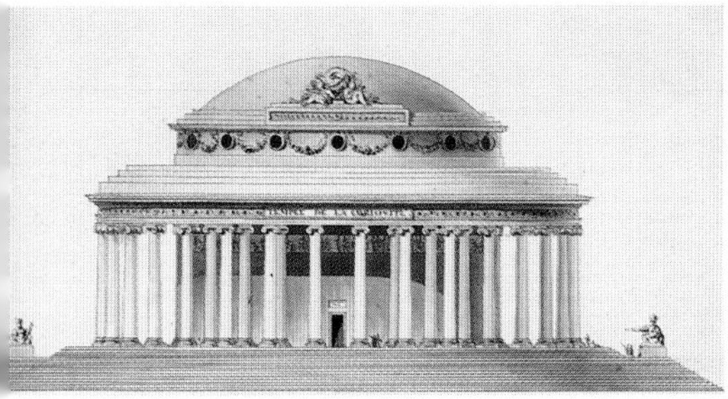

2

3

Worship and celebration, aviation and even the future, are housed in these vaulted spaces—the simple shape of a circle retooled for each intent.

4

1 Based on an illustration in Andrea Pozzo's important treatise on perspective from 1693, this drawing of the interior of a church cupola from the early 18th century dazzles the eye with the complexities of its illusionistic architecture. 2 This 18th-century drawing for a Temple of Curiosity is associated with the work of neo-classical architect Etienne-Louis Boullée, who argued for an architecture of reason that also inspired grandeur and emotion. 3 For the 1939 New York World's Fair, William Lescaze and J. Gordon Carr designed the aviation exhibition building, a hangar-like structure that echoed the form of an airplane. Hugh Ferriss' 1937 drawing, with its dramatic contrasts of light and dark, conveys the architects' confident embrace of technology. 4 French architect Félix Duban designed this ornamental panel as part of the decorations for the 1835 July Festival celebrations, a patriotic three-day holiday in Paris that occurred on the anniversary of the 1830 revolution. 5 Staircase from the château at Chambord, France, as illustrated in a 1715 English edition of Renaissance architect Andrea Palladio's *I Quattro Libri dell'Architettura* (Four Books of Architecture), first published in Italy in 1570.

5

1

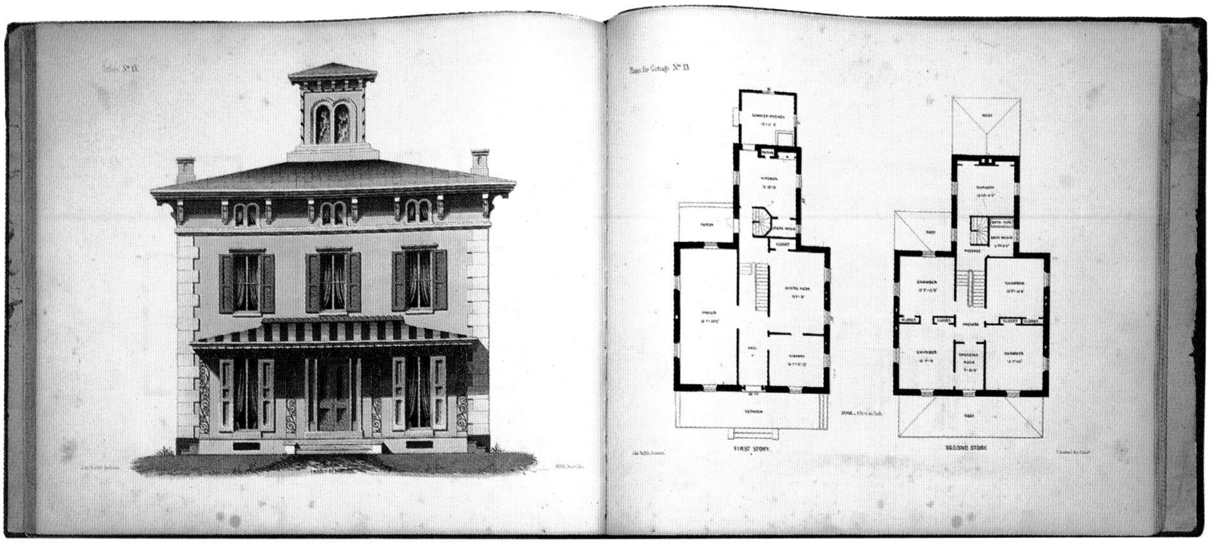

2

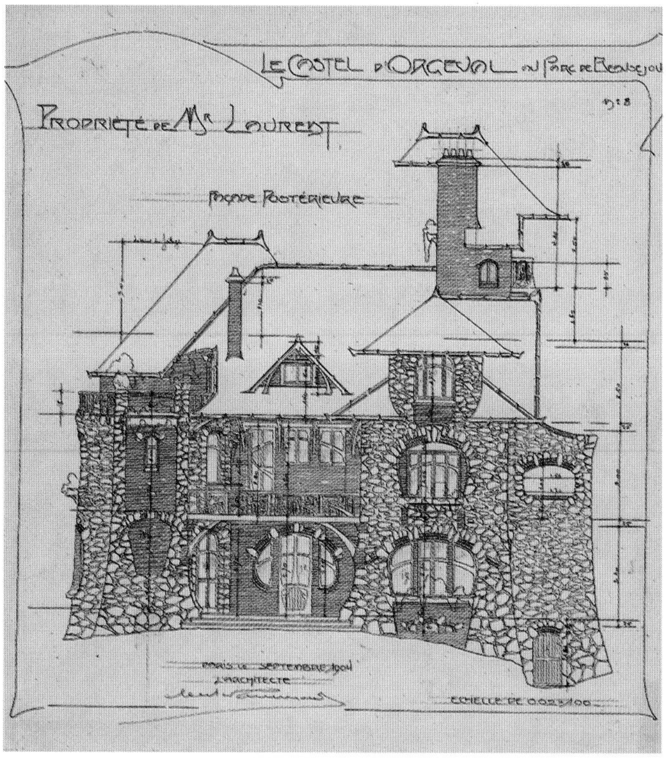

3

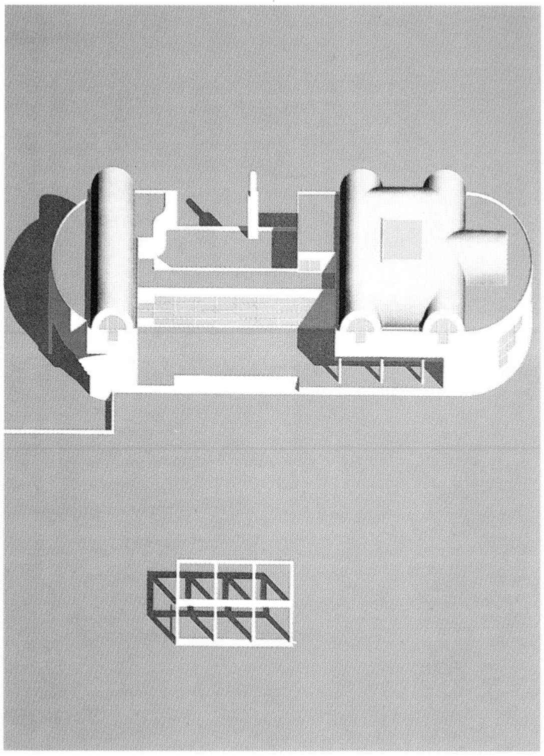

1 A page from an American architectural pattern book from 1861 by Philadelphia architect John Riddell with elevations and house plans for country residences. 2 This 1904 drawing of the Castel D'Orgeval near Paris shows Hector Guimard to be a master of the Art Nouveau style, which is characterized by organic, sinuous forms. The drawing is one of many that would have been used by the builders to construct the house. 3 This presentation print for a "House for Two Generations" by Arata Isozaki is characteristic of his work from the 1970s. Its barrel vaults, grids, and arches reflect both his interest in Italian Renaissance Palladian villas and the modernist architecture of Le Corbusier.

4

4 Donald Deskey first introduced his design for a prefabricated weekend cabin, the "Sportshack," at the 1939-40 New York World's Fair. This lively drawing shows Deskey's attempts to overcome the public's aversion to factory-built homes by using open spaces, new materials, and practical decor.

The idea of a "dream house" haunts the psyche, each generation re-forming the scheme to accommodate new ideas of status, comfort, convenience, and need.

1 A. W. N. Pugin designed these brass door plates for London's new Houses of Parliament, c. 1845. 2 Iron door plate by Louis Sullivan for the Guaranty Building in Buffalo, New York, c. 1894. 3 In 1767, Jacques-Denis Antoine received the commission to design the new French Mint from Louis XV, whose crossed double-L monogram forms part of the grillwork of its gates, featured in this drawing from 1776.

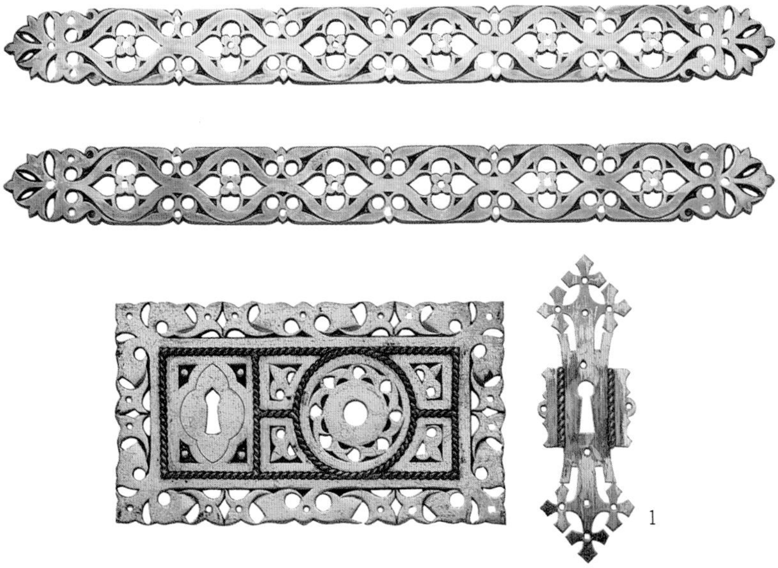

1

2

3

4

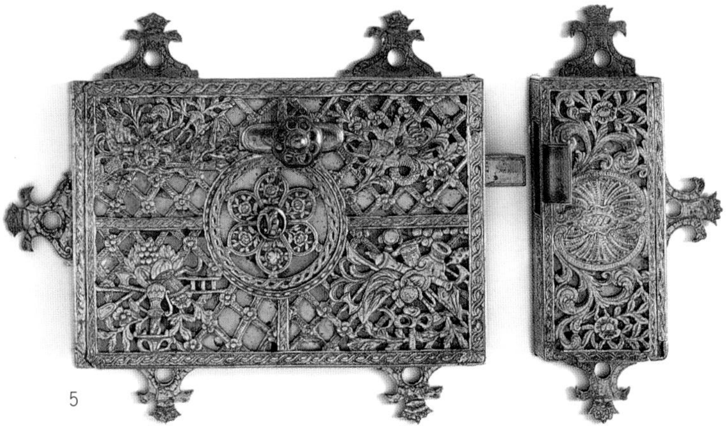

5

4 Once installed at the entrance of the executive suite on the fifty-second floor of the Chanin Building in New York City, these iron and bronze Art Deco gates were designed by René Chambellan in 1928. Ornamental motifs of cog wheels, lightning bolts, and stacks of coins were designed to represent the triumph of commerce and industry. 5 A 17th-century Spanish steel lock.

Gates that advertise wealth and power need locks and bolts to make them secure.

Movable screens and draped cloth create flexible, versatile rooms with elegant economy.

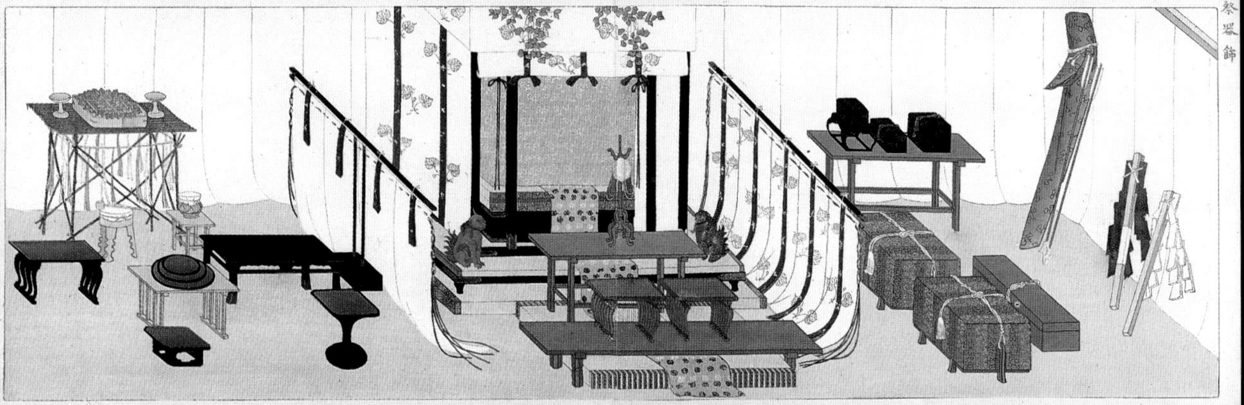

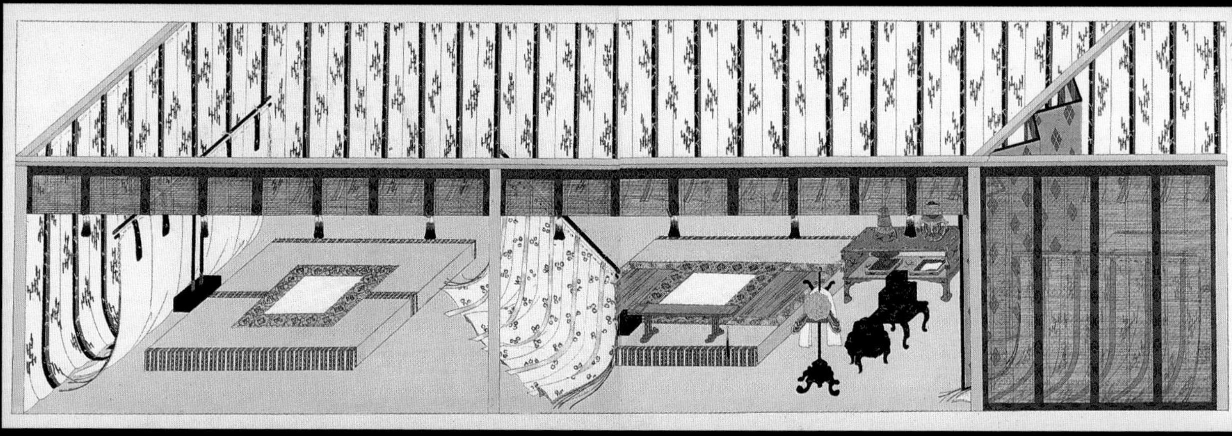

Three plates from *Sixteen Japanese Ceremonies*, an exquisite book first

No people inhabit the interiors, but the necessary utensils are carefully

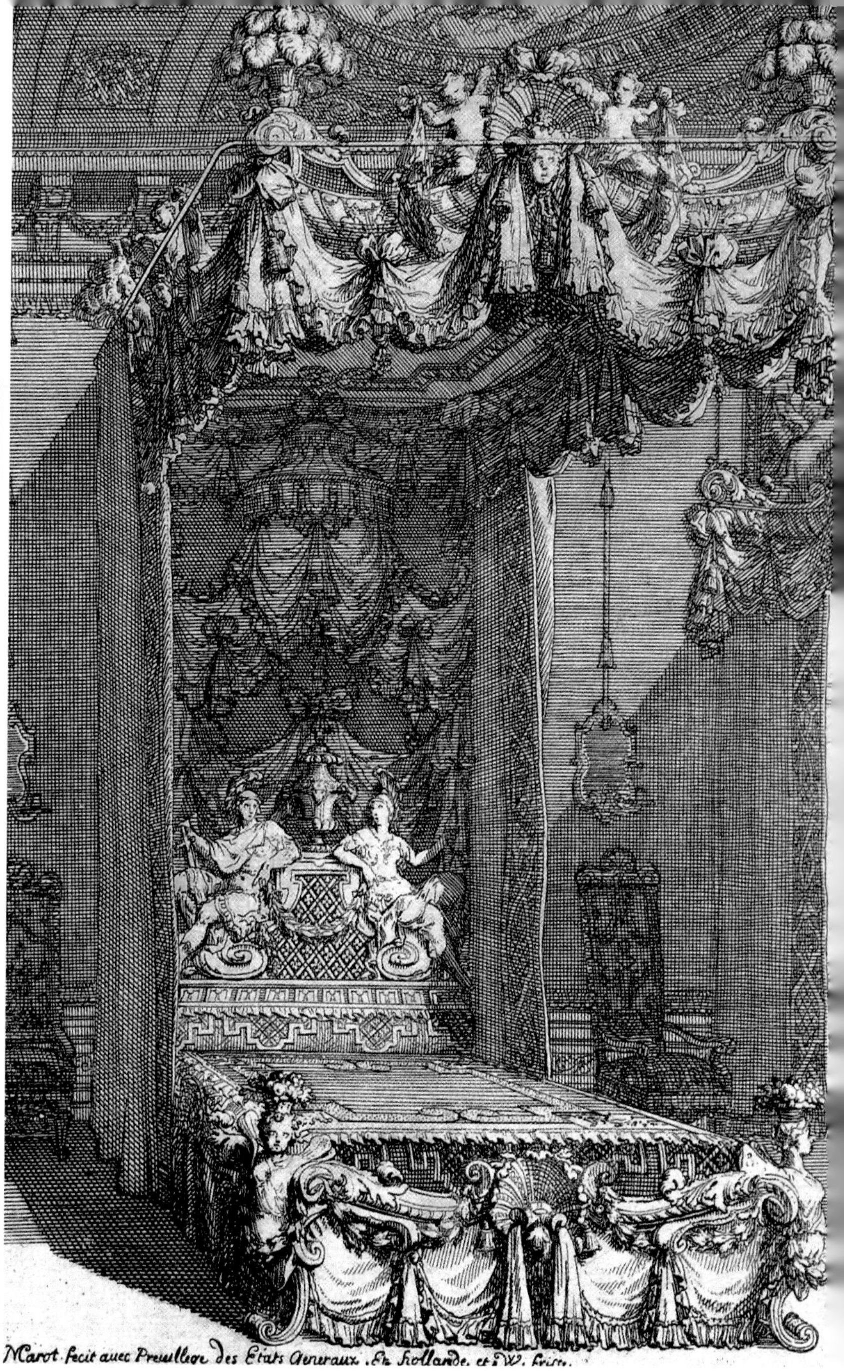

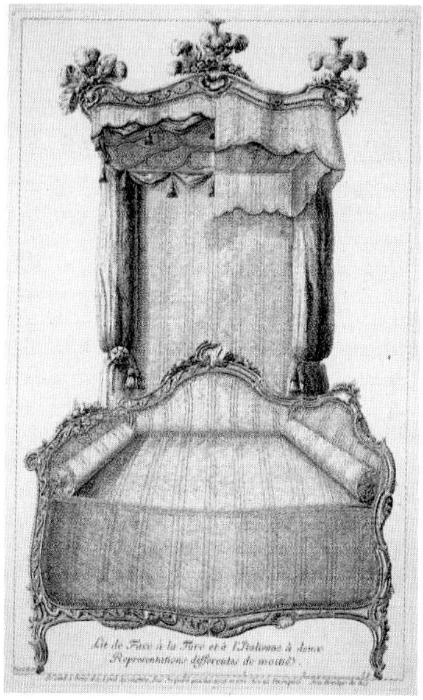

1 Between 1495 and 1503, German printmaker Israhel van Meckenem made a series of engravings depicting scenes of domestic life, including this rare glimpse of a 15th-century bedroom. The realistically detailed setting enhances this captivating portrayal of a couple in the beginning stages of an intimate encounter. 2 A print, c. 1702, of a design for a state bedchamber by Daniel Marot, famed French Huguenot designer. This sumptuous bed would have stood over 15 feet high, proclaiming the wealth and status of its owner.

3 This 18th-century print by Matthew Liard shows a hybrid bed designed in the "Turkish" style on the left and in the "Italian" style on the right, to provide as many choices as possible to potential clients interested in commissioning such a piece.

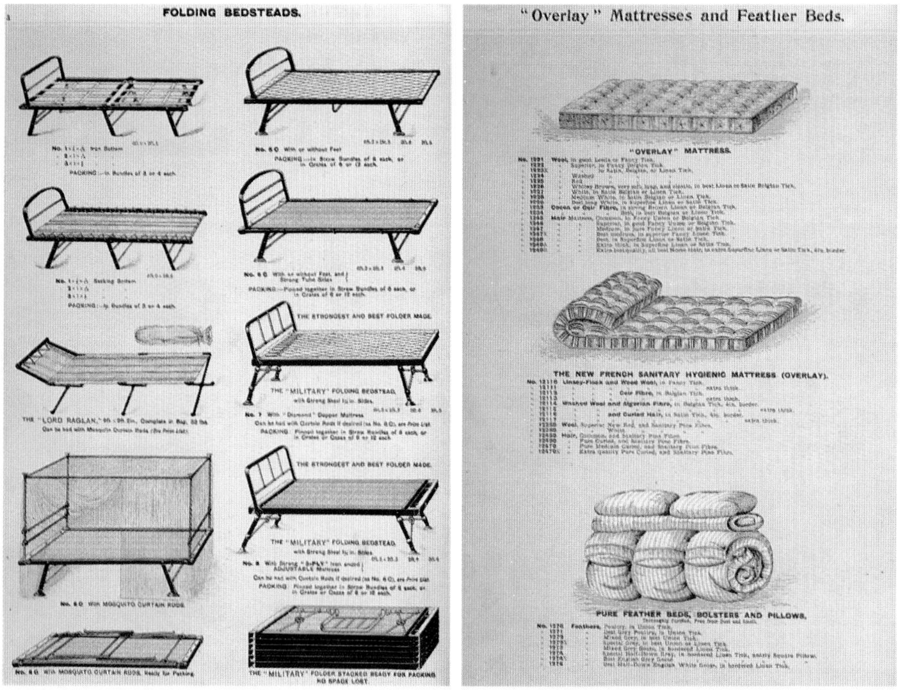

4

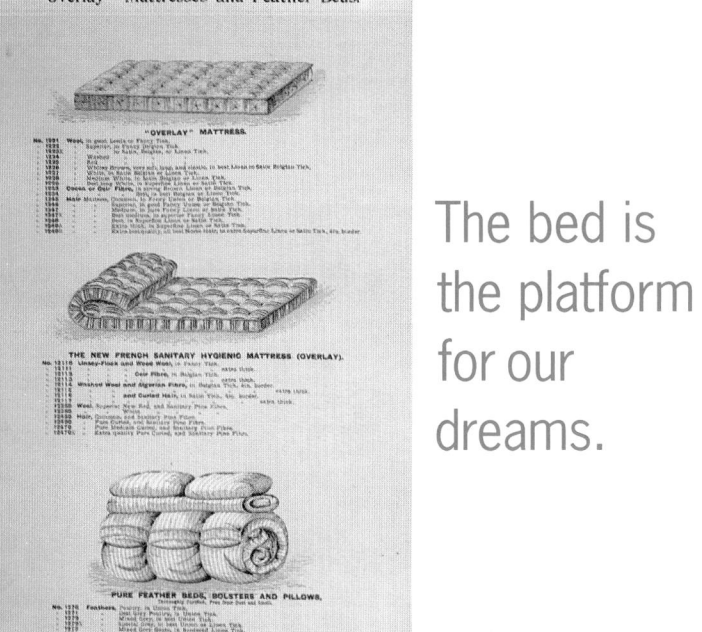

The bed is
the platform
for our
dreams.

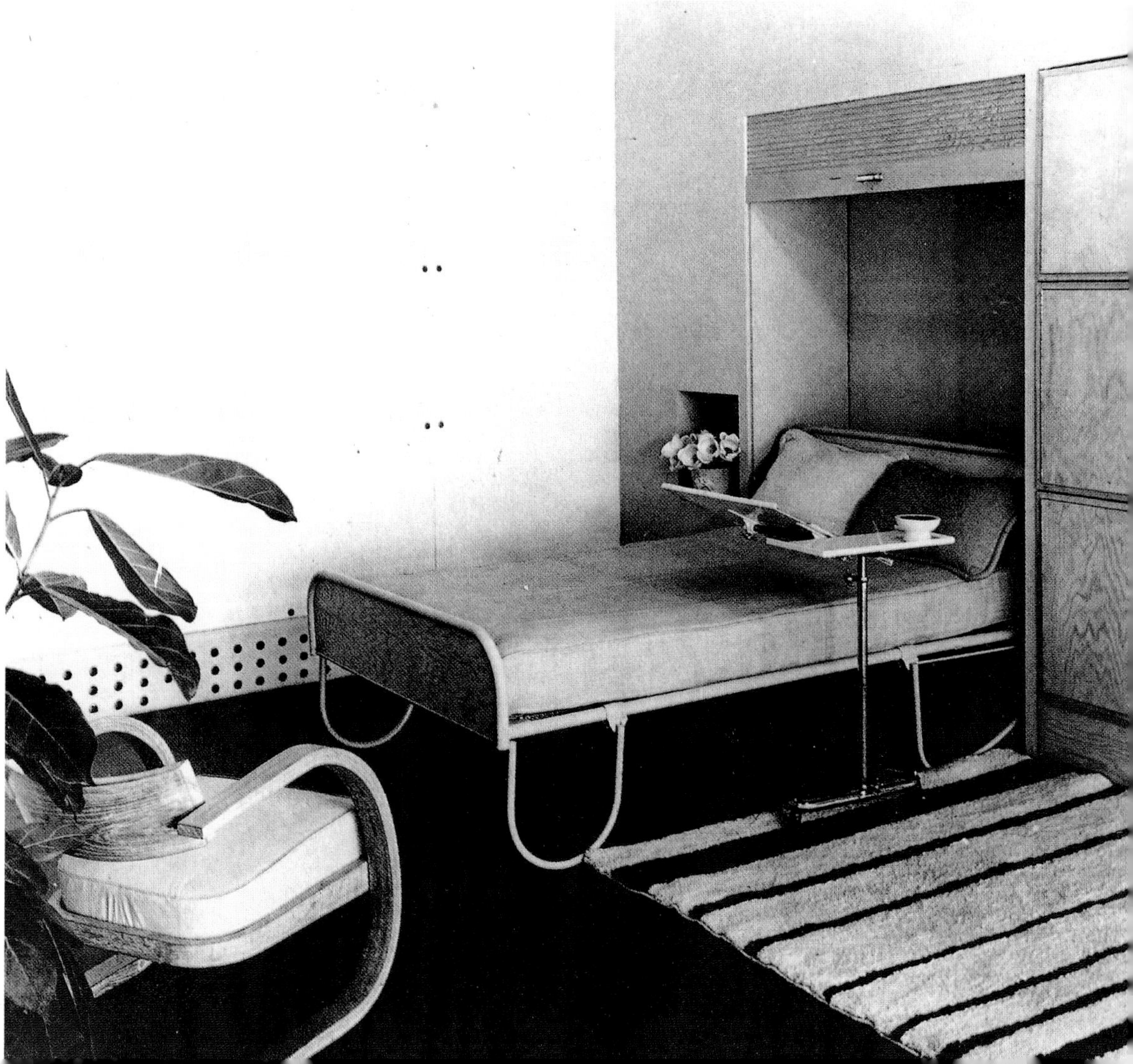

5

4 A Fitter Brothers trade catalogue, c. 1906, advertised its bedsteads to a mass audience. 5 The modern bed disappears altogether with the invention of the "Murphy" bed, such as this French example, in a bedroom designed by Maurice Barret, c. 1937.

1

2

1 Watercolor of apartment of Queen Elizabeth of Prussia, Charlottenburg Palace, Berlin, done in 1864 by Elizabeth Pochhammer. 2 This 1930s drawing by André-Léon Arbus for a luxurious salon reflected his belief that design should resist the machine aesthetic in favor of elegant craftsmanship. 3 Details from the book *Ein Wohnhaus* (A Residential Building) by German architect Bruno Taut. This 1927 treatise documents the liberating use of color characteristic of Taut's later architectural practice.

3

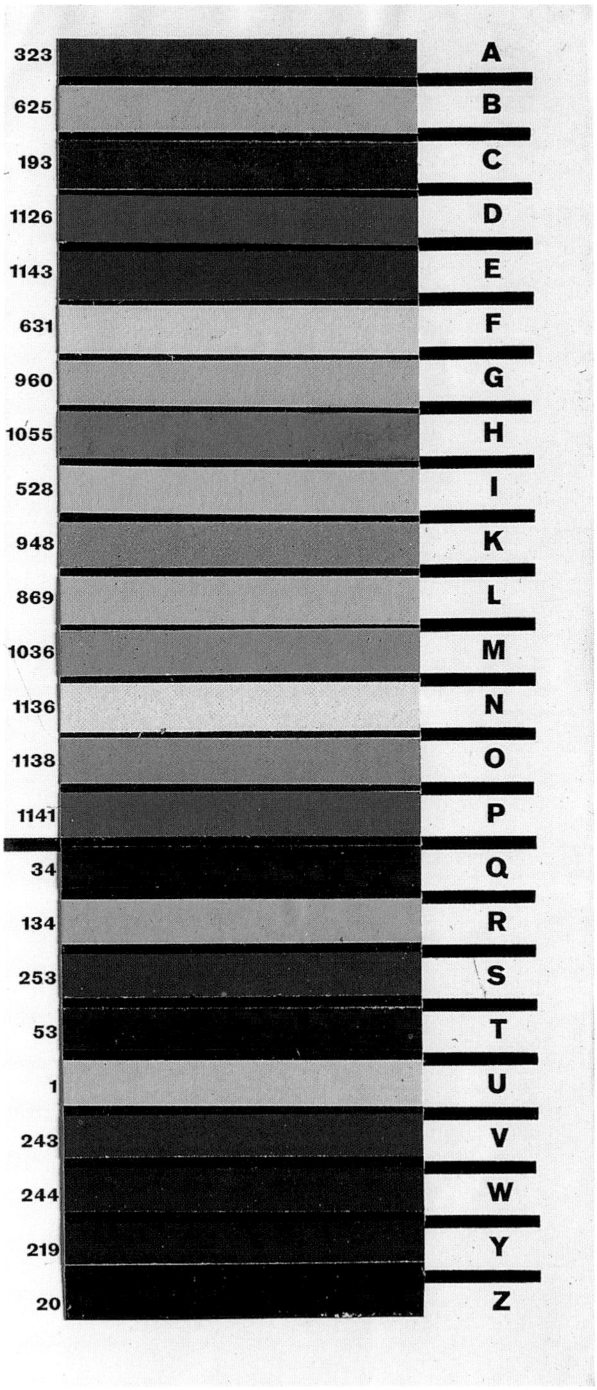

Color gives us the first clue about the personality of a room—and the nature of its inhabitants.

1

Above: 1 This 1830 drawing of a marble mantelpiece by Roman architect Pietro Camporese the Younger emphasizes both the practical and aesthetic aspects of the decorative surround. He may have used it as a presentation drawing for potential clients. 2 The great success of Henry Dreyfuss's Honeywell T86 thermostat can be measured by the millions of American homes in which it was installed since its introduction to the market in 1953. Unlike its rectangular predecessors, its innovative round shape never looks askew on a wall. 3 New methods of producing commercially viable tempered glass inspired René Coulon to design this elegant radiator in 1937.

A space becomes a room when it has warmth and light.

2

1

3

4

5

1 Oil lamp, Italy, 1st century B.C. 2 Drawing for two chandeliers by François-Joseph Belanger, for the Pavilion de Bagatelle, Paris, France, 1777. 3 Candlestick by Hood and Hood, England, 1878-79. 4 "Dragonfly" light, Tiffany Studios, United States, 1900-10. 5 Hanging light, Jac van den Bosch, The Netherlands, 1902. 6 Standing light, Edgar Brandt, France, c. 1925. 7 "Falkland" hanging light, Bruno Munari, Italy, 1964. 8 "Artichoke" hanging light by Poul Henningsen, Denmark, 1958. 9 "Chrome Blender" table light, made by Virginia Restemeyer, United States, 1989. 10 Table light, France, 1980s. 11 Flashlights: Rayovac "Vidor," "Garrity," and Eveready "Compact Industrial," United States, 1980s.

Lighting up the dark is an ageless human endeavor.

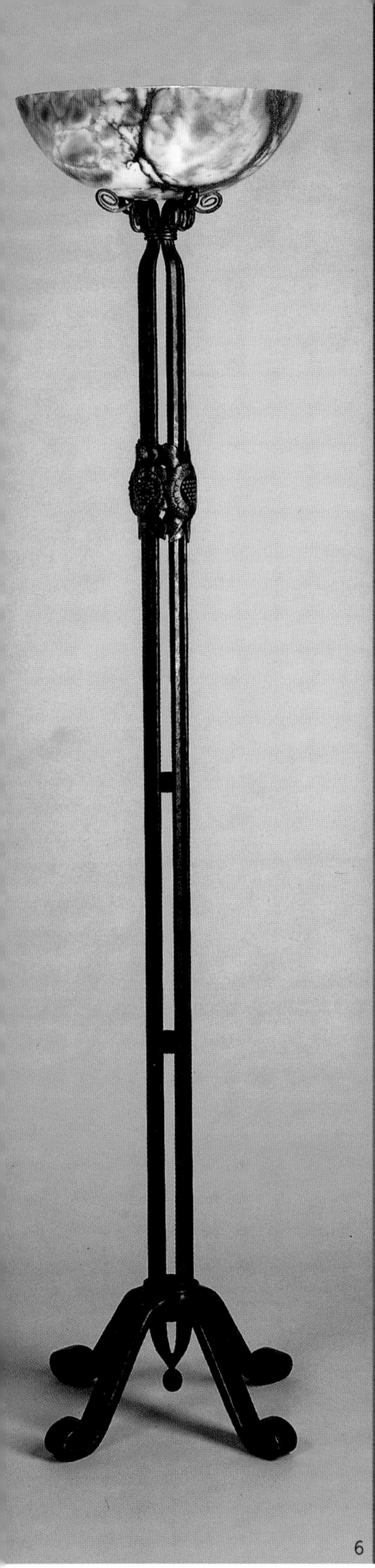

6

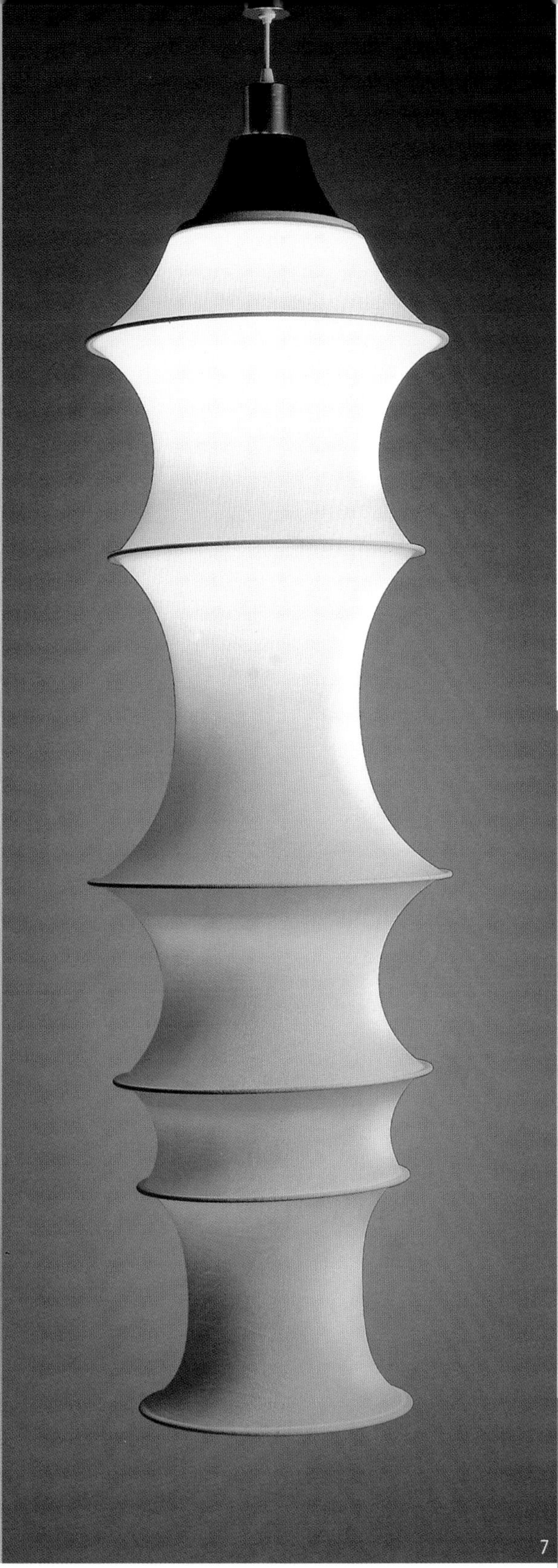

7

8

9

10

11

2

3

1

1 This magnificent cabinet was made at the Sèvres Porcelain Factory in 1824-26 and presented as a French gift of state. It is distinctively clad with large, exquisitely decorated porcelain panels framed in gilt bronze, shown here in detail. 2 Frederick Crace's drawing of a sofa in an alcove with a tented ceiling was probably done for the Prince of Wales' bedroom at the Royal Pavilion, Brighton, c. 1801-04. 3 Drawing of two andirons and a sconce by François-Joseph Belanger, 1777, for the Pavilion de Bagatelle, Paris.

Powerful demonstrations in porcelain and gilt emanate royal prestige.

1

Cultures cross-pollinate through exotic motifs, color, and a sophisticated language of ornament.

1 Frederick Crace's design for the west wall of the Music Room of the Royal Pavilion at Brighton exemplifies the fashion for exotic chinoiserie ornament in the early 19th century. 2 This 1786 wallpaper was produced at the Paris factory of Jean-Baptiste Réveillon. 3 A French Empire furnishing fabric (c. 1810-20) designed to upholster a chair with elements that would have been cut to cover the back, seat, and front rail. 4 Plate from Wilhelm Zahn's 1843 illustrated study of Italian classical and mannerist ornament.

2

3

4

2

Classical design imparts a reassuring sense of history and authority through ancient myths and motifs.

1 "The Reconciliation of Venus and Psyche" is one of 12 scenes in this 19th-century scenic wallpaper designed by Merry-Joseph Blondel and Louis Lafitte. It was first produced by Dufour et Compagnie in 1815-16; this edition by Desfossé et Karth was printed from 1,245 wood-blocks in 1923.

2 This unique factory work-book records variations of egg and dart wallpaper borders, as well as other borders imitating archi-tectural elements, offered by Jean-Baptiste Réveillon. The Réveillon factory was one of the premiere 18th-century French wallpaper producers but fell victim to the French Revolution.

1

The panorama was a way to make yourself the center of the world.

A French scenic wallpaper designed in 1848, "*El Dorado*" depicts four major continents—Africa, Asia, America, and Europe—through their native vegetation and architecture. This non-repeating narrative panorama was made to cover all four walls of a room. A technical tour de force, its 24 panels were block-printed with 1,554 carved-wood blocks.

1 Humphry Repton wrote his *Observations on the Theory and Practice of Landscape Gardening* in 1803. The folio features scenes such as this one supplemented by fold-out pages to provide "before" and "after" views to prospective clients.

before

1

We reorganize nature itself to suit our *idea* of nature.

after

2 André and Paul Vera's 1914 design for "A Love Garden" shows the Vera brothers' development of a mathematically ordered garden style derived from both contemporary Cubist and traditional French design, in reaction to free-form English gardens.

3 The imagery on this early 19th-century plate-printed fabric shows people strolling in a zoological garden and may depict an actual zoo in the city of Nantes, France.

2

3

1 Stag hunting is depicted in "*Décor Chasse et Peche*" (Hunting and Fishing Decoration), a French wallpaper from 1839. *Above:* 2 "Acanthus" wallpaper from 1875 by William Morris, founder of the English Arts and Crafts Movement. 3 Edouard Bénédictus designed this cotton and rayon furnishing fabric "*Les Jets d'Eau*" (The Fountains) for the Grand Reception Room of the French embassy's exhibit at the 1925 Paris *Exposition International des Arts Décoratifs et Industriels Modernes*. 4 Traditionally, wallpaper has imitated more expensive finishes. These papers simulate woodgrains and date from the early 19th century through 1992.

1

3

4

Domesticating the outdoors, we cover our walls and surround ourselves with illusions of nature.

1

Flora and fauna are printed and woven into fabrics for walls and furnishings.

Opposite, clockwise from upper left: 2 Stylized flowers create a strong vertical and diagonal sense of movement in this block-printed furnishing fabric designed by Josef Hillerbrand and produced at the Deutsche Werkstätte in 1926. 3 *"Tharrkarre,"* batik hanging, Judith Kngawarreye, Australian, 1989. 4 *"La Terre"* (The Earth) is one of four textiles representing the elements—earth, fire, air, water—designed by Mlle Clarinval and exhibited at the 1925 Paris *Exposition International des Arts Décoratifs et Industriels Modernes.* 5 "Peri," a woven silk furnishing fabric designed in 1870-71 by Owen Jones, author of *The Grammar of Ornament,* demonstrates his conviction that pattern for flat surfaces should be rendered without illusion.

1 Lucienne Day designed the screen-printed furnishing fabric "Calyx" for the 1951 Festival of Britain. Widely published, it became a signature image of the period.

A basic human posture takes on infinite design postures.

1 Armchair, England, c. 1880. 2 Side chair, or *sgabello,* Italy, 16th century. 3 "Sacco" chair, Piero Gatti, Cesare Paolini, Franco Teodoro for Zanotta, Italy, 1969. 4 Child's chair, Charles and Ray Eames for Herman Miller, United States, 1944. 5 Chinese export armchair, c. 1815. 6 Armchair, Warren McArthur, United States, c. 1935. 7 Armchair, Russel Wright, United States, 1934. 8 Armchair, possibly France, c. 1810. 9 Side chair, attributed to Carlo Zen, Italy, c. 1900. 10 "Grasshopper" armchair and ottoman, Eero Saarinen for Knoll, United States, 1945-50. 11 Side chairs, from *Le Garde-meuble,* Vol. 79, France, 1850s. 12 Armchair, Marcel Breuer, Germany, 1925. 13 Pair of side chairs, René Drouet and Madeleine Luka, France, c. 1939. 14 Side chair for Imperial Hotel, Tokyo, Frank Lloyd Wright, c. 1920. 15 Purkersdorf Sanatorium side chair, Josef Hoffmann, Austria, 1904-06. 16 Armchair, Warren Platner for Knoll, United States, 1965. 17 Side chair, Emile Gallé, France, c. 1900. 18 Rocking chair, Charles and Ray Eames for Herman Miller, United States, c. 1950. 19 Side chair, Peter Behrens, Germany, 1902. 20 "Easy Edges" chair, Frank O. Gehry, United States, 1971-72. 21 Corner loveseat, from *Le Garde-meuble,* Vol. 79, France, 1850s.

1

2

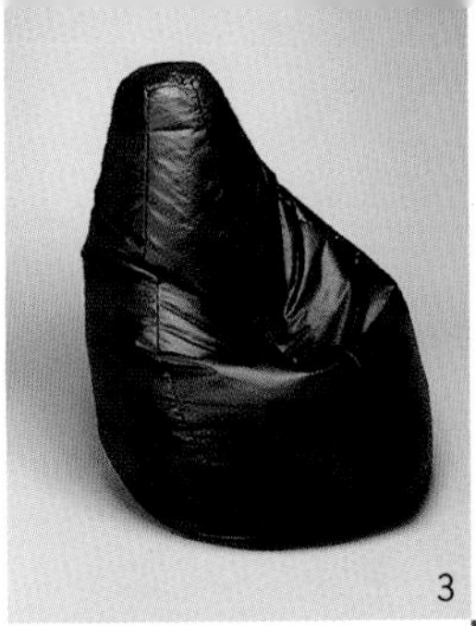
3

4

5

6

7

8

9

10

11

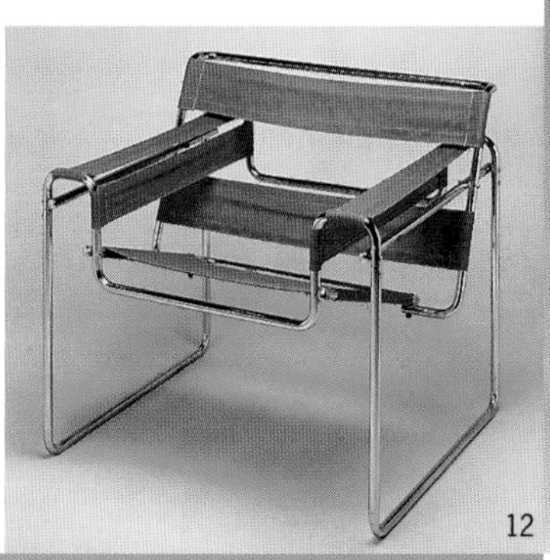
12

13

14

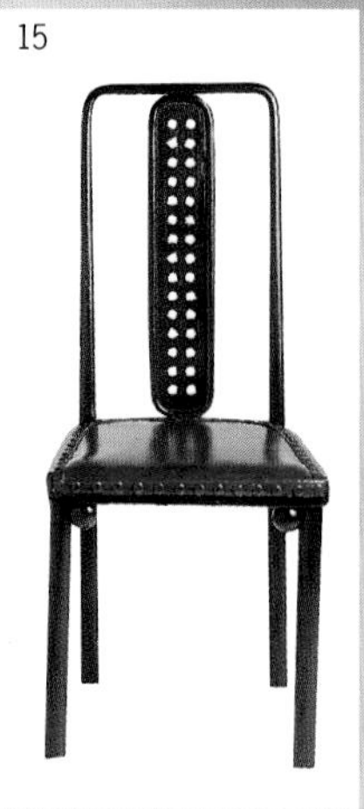
15

16

17

18

19

20

21

Co

Design for mmunicating

This finely crafted pop-up book features lively images from the classic tale *Alice's Adventures in Wonderland.*

Every schoolchild knows that what separates us from the animals is our ability to tell stories: short stories, long stories, sound bites, fables, and epics. Stories appear in books, on posters, in ledgers, on packages, billboards, and wallpaper. They are central to our understanding of ourselves and how we come to terms with the events and people with whom our lives intersect. To tell those stories, designers have merged sophisticated systems of symbols—alphabets, hieroglyphs, ideograms, and pictographs—with the powerful language of images.

The alphabet is an image in and of itself, likely to conjure the sing-song rhyme of one of our earliest feats of memory and recollections of classroom exercises in the art of penmanship. Upper- and lower-case letters have filled volumes of copybooks and have been spaced across the tops of countless chalkboards. In centuries past, when girls were largely absent from the schoolroom, letters of the alphabet were also sewn as outlines on cloth samplers. More often than not, literacy in the art of embroidery was the primary objective of the lesson, a cornerstone of a well-born girl's education in the 18th and 19th centuries. The Museum's unique collection of embroidered samplers shows how young women in Europe and America disciplined their needlework to conform to the architecture of the alphabet.

In 1773, ten-year old Catharine Congdon of Newport, Rhode Island, embroidered this sampler as part of her schooling.

As much as we are fascinated by the idiosyncracies of handwriting—be it sewn or scripted—we have also sought to standardize it. Typographers' manuals are filled with letterforms that both satisfy our need for legibility and entertain our fascination with invention. Partly to distinguish one from the other, and partly to convey their character, we call different families of letters type*faces*. And we name them. For example, the text of this essay is set in a classicized font, or typeface, called Sabon and is offset by headers and captions in modern News Gothic to reflect the contemporary relevance of a museum collection rich in history.

The choice of a typeface is often our first clue to the content of the story being told. Before the modern field of graphic design existed, printers guided clients in selecting typefaces to suit their texts. A 19th-century printer's catalogue such as the Museum's 1855 book produced by James Conner & Sons shows the period's passion for expressive alphabets. In the selection offered by this New York shop, image and language converge in letters made of logs that spell out "huge water oak and pine," while the phrase "Rossini's Barber of Seville" advertises type patterned with the diagonal stripes of a barbershop pole.

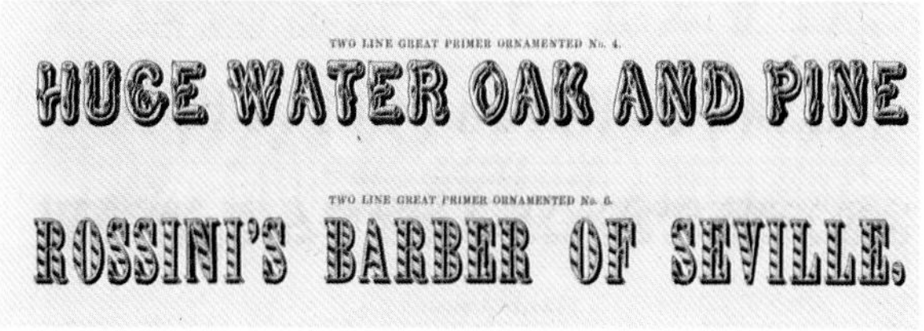

Type samples from James Conner & Sons' printer's manual, 1855.

Rejecting the obvious sentimentality of the past, the avant-garde of the 1910s and '20s also understood that typography was a visual language unto itself, capable of affecting the meaning of what we read. The Italian Futurists played with spacing and letterforms to convey their obsession with the speed and velocity of the new century's machines and weapons. Russian Constructivists telegraphed the propaganda of the new Soviet Republic to largely illiterate masses with dramatically scaled letters layered over photo montages of industrious workers. Language was

The visual message of Valentina Kulagina's 1930 International Women Workers Day poster is clear: women are vital in the work force and can help weave a strong social fabric for the Soviet future.

manipulated by the structure of the page, opening the door for subsequent generations' experiments with the way people read.

The expressionism that dominated early 20th-century vanguard movements was mitigated by another equally powerful response to fin-de-siècle change: the urge to rationalize, to make sense of technology's potential for improving daily life. It is from these two streams that graphic design continues to draw its energy today. In the 1930s and '40s, this emerging profession was dominated by an aesthetic that married the abstraction of the avant-garde with the idea of functionalism. Geometry was thought to offer a series of first principles: the cleaner the form, the clearer its meaning and use would be. Architects, artists, industrial designers, and commercial illustrators all commingled, and professional distinctions blurred in the quest for a universal language of design.

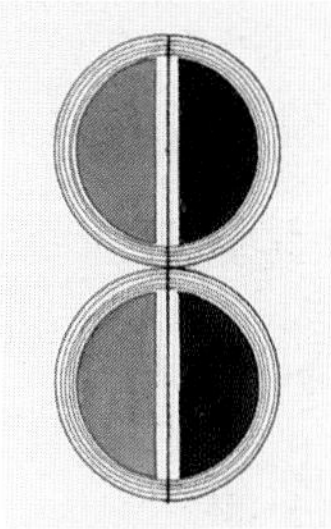

Three monograms, "B.E.," "E.L.," and "N.S.," by Jean Puiforcat.

Jean Puiforcat was one such designer. Known for his elegant Art Deco silver tablewares, furniture, jewelry, and glassware, Puiforcat also applied his interest in geometric forms to the design of a series of monograms. The Museum's colorful renderings, done between 1930 and 1945, show configurations of pairs of initials that may have embellished flatware, jewelry, toiletry sets, or stationery. Highly stylized and personal, Puiforcat's monograms are akin to private signatures, and yet they are not unlike logos, the symbols of the public persona that graphic designers create for corporate entities.

One of the most interesting examples of design in the service of industry can be found in the Museum's archives of Czech-born designer Ladislav Sutnar. Like other designers, Sutnar had a multifaceted career.

He was a painter and a designer of products, exhibitions, and graphics. Shortly after emigrating to the U.S. in 1939, he applied his prodigious talent as an information designer to the organization of data for the architectural resource publication *Sweet's Catalogue Service*. In designing for *Sweet's*, Sutnar applied the modernist dictum of simplicity to the task of illustrating the complex procedures and equipment used in fabricating buildings and their infrastructure. Sutnar's elegant diagrams and straightforward typography transformed this workaday reference tool into a paradigm of design that was at once functional and expressive.

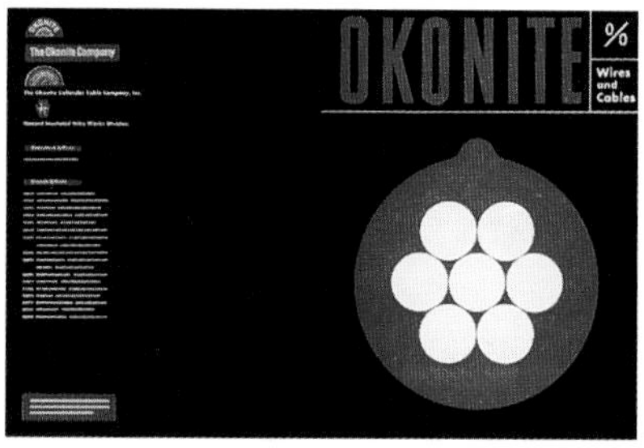

Detail of a page from *Catalog Design* by Ladislav Sutnar and his partner Knud Lönberg-Holm, 1944.

Critical knowledge—industrial secrets passed from master to apprentice, from inventor to producer—can be found between the covers of instruction manuals. Nineteenth-century textile printers, like their counterparts in book publishing, were the purveyors of choice for their clientele. Instead of displaying typefaces, they offered color selections and kept record books with highly prized formulas for achieving color-fast dyes. Textile research today is aided by scrapbooks such as the one assembled by the English dyer Thomas Ratcliffe between 1812 and 1822. In it, small swatches of fabric are annotated with dyeing recipes drawn from his own work as well as that of his competitors.

Detail from a page of swatches from Thomas Ratcliffe's scrapbook, 1812-22.

Among the most famous artifacts in the genre of instruction is French philosopher Denis Diderot's *Encyclopédie, ou Dictionnaire raisonné des sciences, des arts, et des métiers* (*Encyclopedia, or Methodical Dictionary of the Sciences, Arts, and Trades*), produced over the course of fourteen years from 1751 to 1765. The Museum owns fourteen of the twenty-seven volumes of this seminal work of the Age of Enlightenment, including the especially valuable sections devoted to the craft industries of France prior to the Revolution. In undertaking the task of documenting the basic principles of every art and science, Diderot sought to privilege knowledge and reason over the reactionary forces of church and state in 18th-century France. Bridging the realms of ideas and documentation, the *Encyclopédie* was both a work of instruction and propaganda.

Illustration of an upholsterer at work from Denis Diderot's *Encyclopédie*, 1751-65.

The art of persuasion comes in many guises—education and entertainment chief among them—and can be exercised in almost any medium. One of the more novel examples can be found in the Museum's collection of Soviet porcelains. Part of Lenin's 1918 Plan for Monumental Propaganda included enlisting the porcelain factories of the czars into the service of the state. A cache of plates and tablewares, originally cast and fired for the imperial court—but left undecorated in the wake of civil war—was co-opted to promote the cause of the Revolution. This unique form of propaganda yielded a profusion of colorful plates and vessels in styles that borrowed from the traditions of folk art and the brave new imagery of the avant-garde. Hammers and sickles were encircled by stylized ribbons; wheat and harvesting tools were nestled in fields of flowers and Cyrillic script; factory and farm workers were surrounded by the fruits of their collective labors.

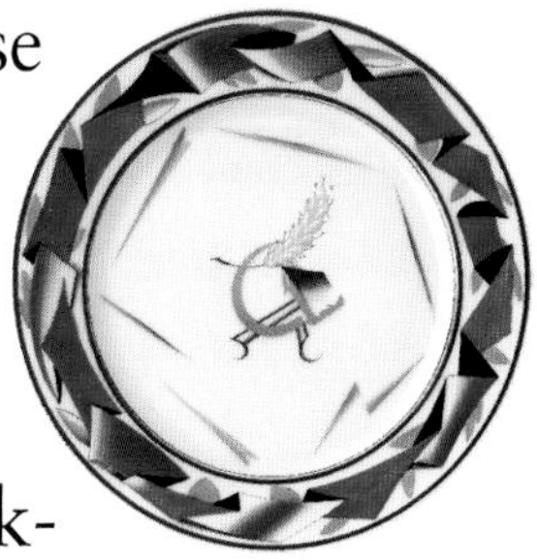

Soviet propaganda porcelain plate by Sergei Chekhonin, 1919.

New regimes have historically depended on designers to bolster their political agendas. In 1789 the French Revolution deposed the rule

Detail of French Revolutionary wallpaper, c. 1792.

of monarchy and proclaimed the rights of the common man. The leadership of the revolutionary government quickly replaced symbols of royalty with republican icons, even commissioning wall-paper to decorate law courts, administrative offices, and other public buildings. The Museum's collection features one of these rare papers, which is patterned with the red Phrygian caps of revolutionary patriots; red, white, and blue cockades displaying the colors of the new state; and Roman fasces (an ax bundled with rods) and oak wreaths that conferred the legitimacy of history on the new Republic.

While Soviet porcelains and revolutionary wallpapers have lost the urgency of their messages, they retain their appeal and value as records of history. Because time quickly overshadows yesterday's events, commemoratives are at their essence artifacts of memory. Modest accessories such as scarves and handkerchiefs tell stories that are part of a nation's folklore. In 1936 the world was scandalized when England's King Edward VIII gave up his throne to marry the American divorcée Wallace Simpson. Souvenir handkerchiefs that featured Edward's portrait were hastily recreated with the name and picture of George VI instead. The few printed with Edward's portrait before his abdication are now rare collectors' items.

Coronation scarf from 1936, featuring Edward VIII.

Heads of state were not only the subjects of souvenirs; they were also given them. The Museum's collection includes an unmounted silk fan leaf presented to Napoleon's wife Josephine de Beauharnais by the

"The Grand Booby" from Mark Catesby's *Natural History...*, 1771.

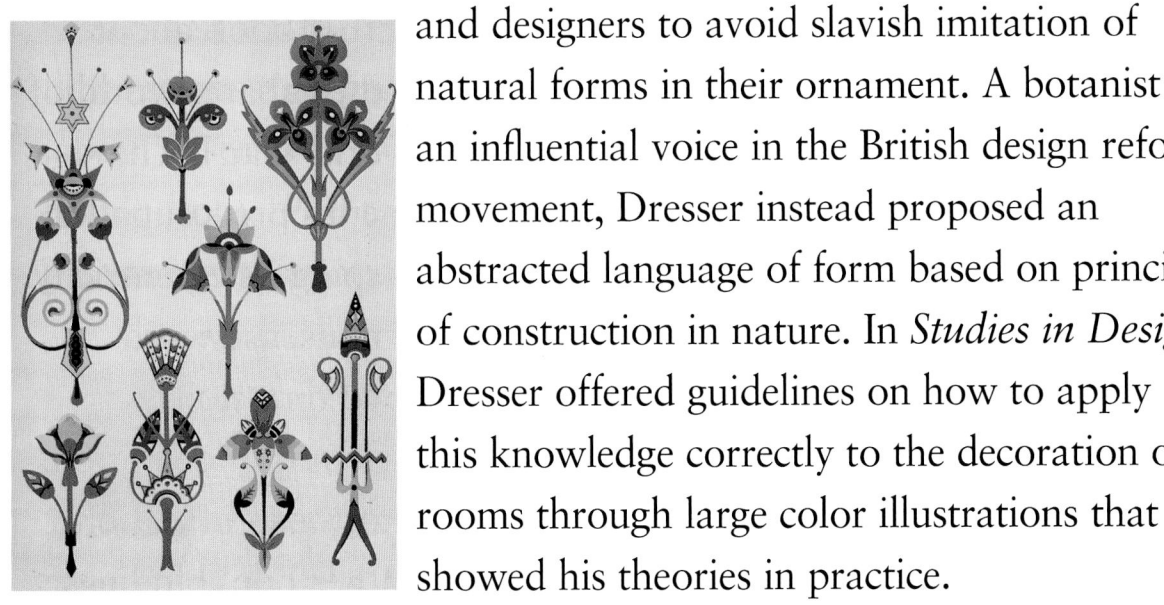

sometimes encrusted with gemstones and gilded with ornamental calligraphy, announce the stature of their owner and the preciousness of the printed word. In some books, words are entirely superseded by the extraordinary quality of the images they describe. *The Natural History of Carolina, Florida and the Bahama Islands* features brilliantly colored plates of the birds, plants, and animals observed by English scientist Mark Catesby on his travels in the Americas from 1683 to 1749. So appealing were his illustrations that the book became a common resource for fabric, porcelain, and wallcovering designers of the day.

"Le Sifilet" ("Singing Bird") from Francois Le Vaillant's *Histoire naturelle des oiseaux de paradis...*, 1806.

Designers themselves have been among the most prolific sources of ideas and inspiration for their fellow practitioners. The 19th-century English designer Christopher Dresser styled himself as an "art advisor." Prescriptives such as his 1876 *Studies in Design* advised architects and designers to avoid slavish imitation of natural forms in their ornament. A botanist and an influential voice in the British design reform movement, Dresser instead proposed an abstracted language of form based on principles of construction in nature. In *Studies in Design*, Dresser offered guidelines on how to apply this knowledge correctly to the decoration of rooms through large color illustrations that showed his theories in practice.

Detail from Christopher Dresser's *Studies in Design*, 1876.

Whether didactic or discrete, designers communicate the spirit of their time—be it a classical faith in reason and order, or a romantic disposition toward the mysterious and personal. At the same time, they are charged with communicating a wealth of practical information about how the material world works and could work better for us. All communication requires a form and a vehicle. It is the designer's task to both code and decode the messages we send.

This book is such a vehicle. *Design for Life* is simultaneously a compendium of the National Design Museum's collections and an argument for its mission: to situate design in the realm of daily life. It has been designed to celebrate the collection's extraordinary diversity and clarify the commonality of purpose of the objects made to serve us, regardless of culture or time period. In the end, this book is also an object introduced into the world, and it too will be seen as particular to our time, at the close of a century, on the cusp of a new millennium.

2

1 Precious tooled, embossed, and bejeweled bookbindings from the Museum's rare book collection. *Top to bottom: Wine, Women, and Song...*, 1884; *The Durbar*, a book on the people of Delhi, India, 1903; *Andächtiges Geist- und Trostreiches Gebett-Büchlein* (Prayer Book for Solace and Spiritual Enrichment), 1675.
2 These bookplates from 1929 are illustrative of Rockwell Kent's powerful, dramatic work.

Noblemen would personalize their libraries by having their books extravagantly bound in leather and embossed with gold. Libraries were within the reach of the masses by the 19th century, and the stamp of ownership continued with printed bookplates.

1

The book becomes a metaphor for the nourishment of mind and body.

1

1 Huntley & Palmer's biscuit tin in the form of a stack of books, England, 1901. 2 Photograph of the storefront of Brodard & Taupin, printer and bookbinder, Paris, c. 1924.

2

BRODARD & TAUPIN
FABRICATION MÉCANIQUE DU LIVRE
Spécialité de Carnets et Agendas de publicité.
EN GRANDES SÉRIES
Fournisseurs des Grands Magasins.
ARD & TAUPIN
BRICATION MÉCANIQUE
U LIVRE
ME I
RIMERIE
BRODARD & TAU
LA FABRICATION MÉC
DU LIVR
TOME
RELIU

1

Quills of birds were used as pens—suggesting an imaginary alliance between writing and flying. Mercury, Roman god of eloquence and travel, is an apt metaphor to represent the post.

2

3

1 This engraving from *L'Art d'Ecrire*, a late 18th-century manual on penmanship, illustrates the importance of posture and good writing tools. 2 *Post Office*, 1936, poster design by Edward McKnight Kauffer for the British General Post Office. 3 William Metzig executed a number of advertising schemes for the Pelikan pen company during the 1930s. The flowing lines of this striking poster create an instant and memorable link between the image and the product.

Pelikan
Pelikan
PATENT

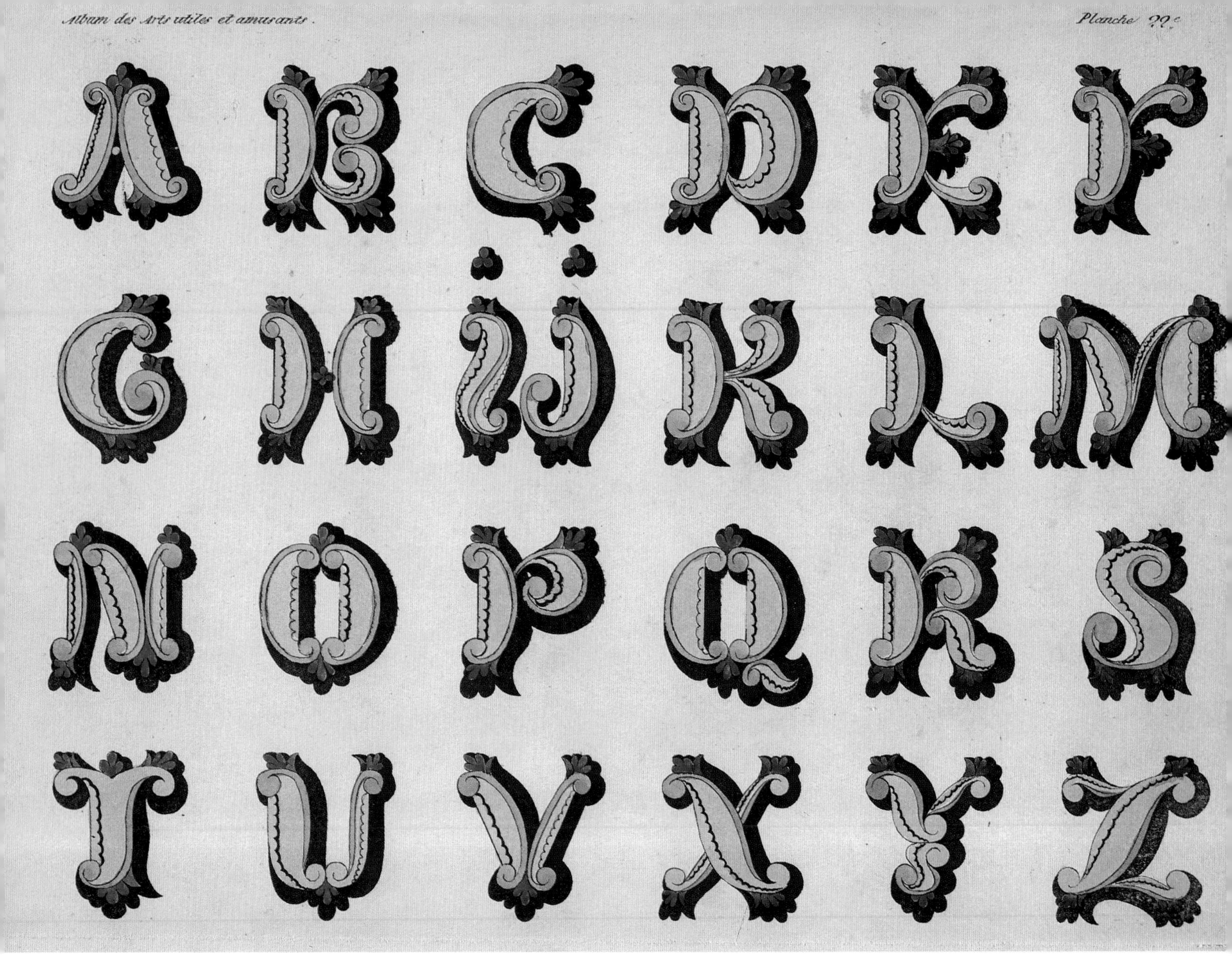

1

2

1 Intended as a "how-to" manual for young people, this French book from 1832 describes various craft hobbies and pastimes that could usefully occupy leisure hours or rainy days at home. Offering only a few examples of decorative lettering, instructions were provided so readers could invent their own novel alphabet styles. 2 Children could learn the alphabet from this Wedgwood porcelain bowl designed by Eric Ravilious and made in England in 1937.

The alphabet itself is brought to life for amusement, instruction, communciation, and commerce.

3

3 For this poster based on a 1951 book jacket design for *The Dada Painters and Poets: An Anthology* (edited by Robert Motherwell), Paul Rand uses overlapping, monumental letterforms to interpret the Dada movement's confrontational use of language and typography. His graphic design for corporations such as IBM brought mid-century modernism to an even broader public and had enormous impact on American business communications.

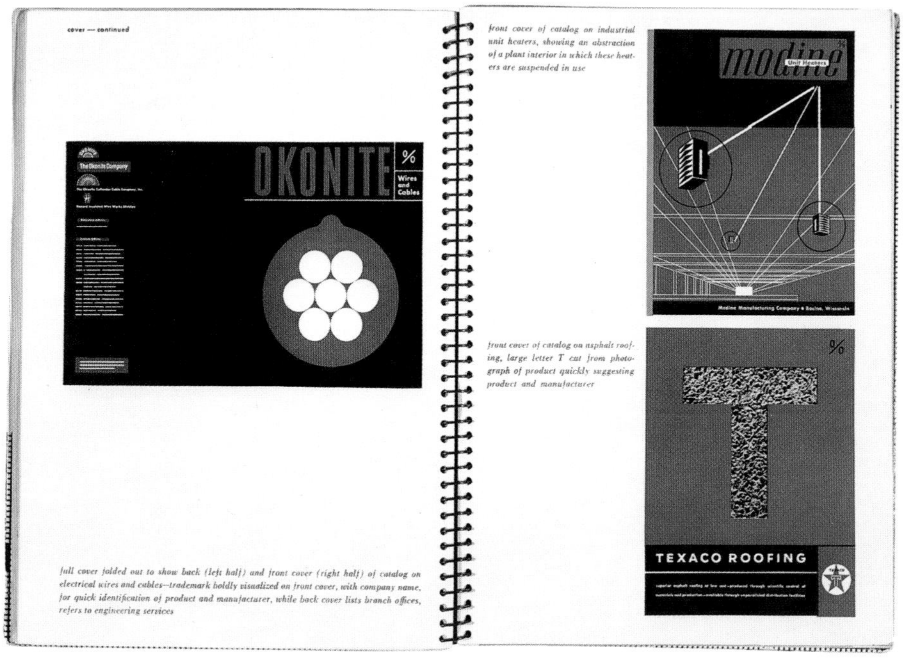

4

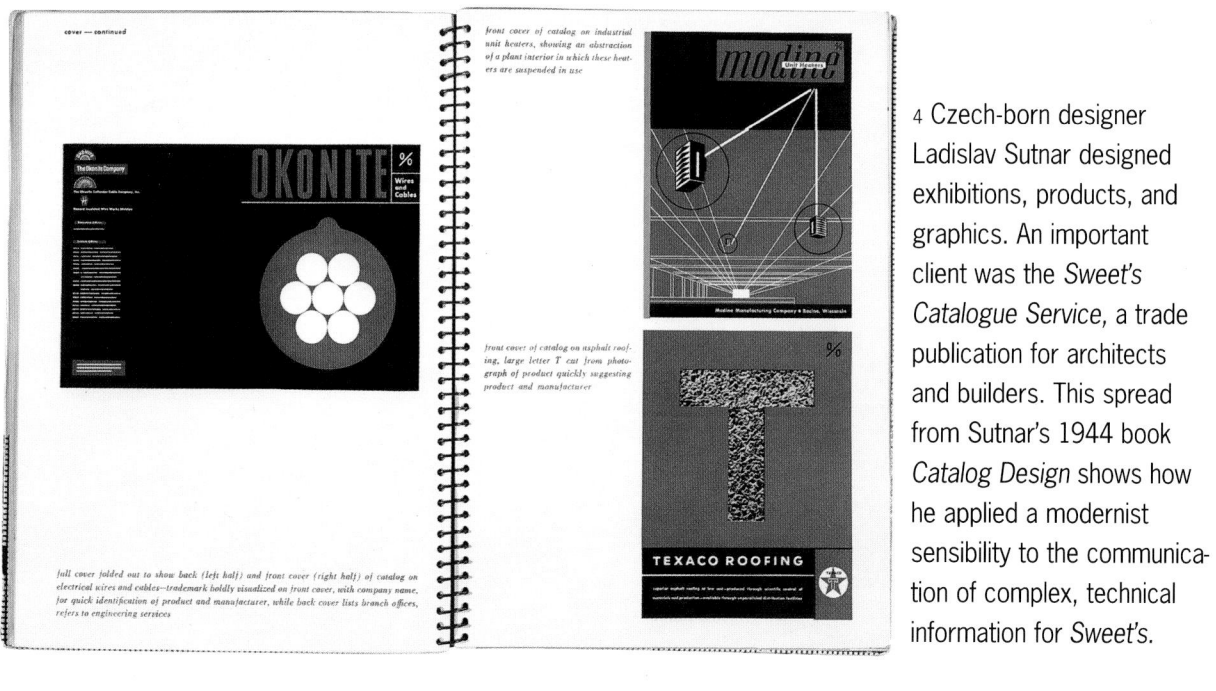

4 Czech-born designer Ladislav Sutnar designed exhibitions, products, and graphics. An important client was the *Sweet's Catalogue Service,* a trade publication for architects and builders. This spread from Sutnar's 1944 book *Catalog Design* shows how he applied a modernist sensibility to the communication of complex, technical information for *Sweet's.*

1

2

3

1 A page from *Alphabete=Alphabets...*, published in the 1870s in Germany, illustrating elaborate, pictorial renderings of the alphabet.
2 Designed in the first half of the 20th century, cigar labels, such as this one for Moro Light, served as highly decorative packaging for wooden cigar boxes.

A world of information is telegraphed by visual shorthand within the outlines of a letter or the confines of a swatch of wallpaper.

3 A 19th-century engraved brass printing plate of the letter "B" (reversed in the printing process) decorated with scenes of Adam and Eve, the apple and serpent, palmettes, and scrolls.
4 Compiled as a production record from 1822 to 1830, this workbook is a rare document of the variety of wallpapers offered by Jacquemart et Bénard. This Parisian firm was active from 1770 to 1840, a period during which French wallpaper was considered technically and aesthetically superior to other European and American production.

4

1

2

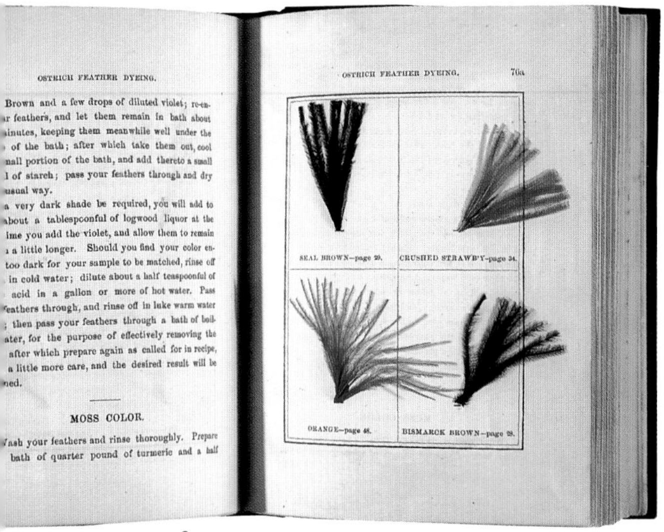

3

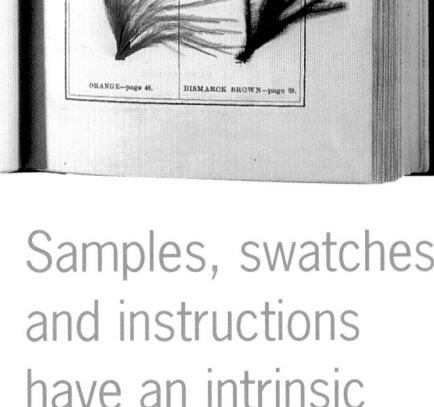

Samples, swatches, and instructions have an intrinsic beauty of their own.

1 An image of an upholsterer at work from French Enlightenment philosopher Denis Diderot's *Encyclopédie* (*Encyclopedia*), 1751-65.
2 The front and back of this French 19th-century plate are painted with 50 different borders and medallions based on Indian, Islamic, and Persian motifs. Sample plates were used to choose patterns for matching table services.
3 An unusual "how-to" book, *The Practical Ostrich Feather Dyer,* was written by Alexander Paul in 1888.
4 Design for a textile, with possible variations, by Louis-Albert DuBois, Switzerland, c. 1801.

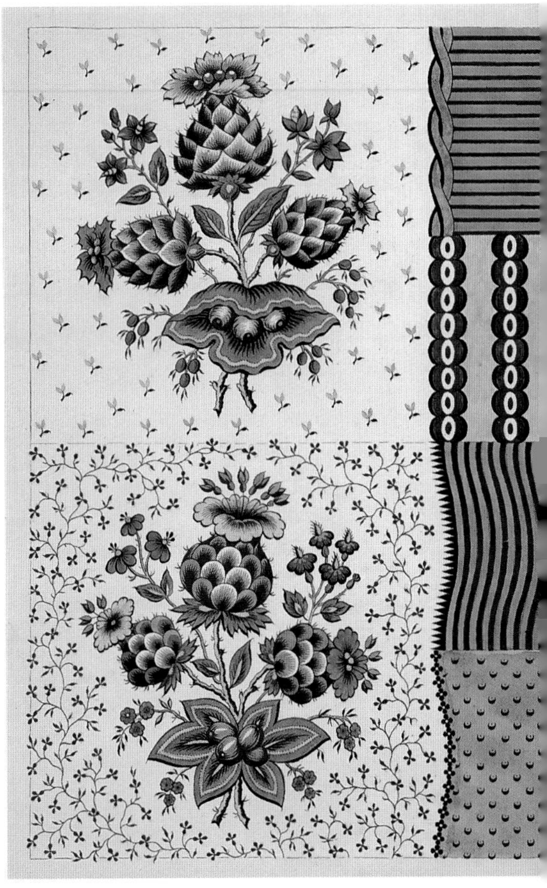

4

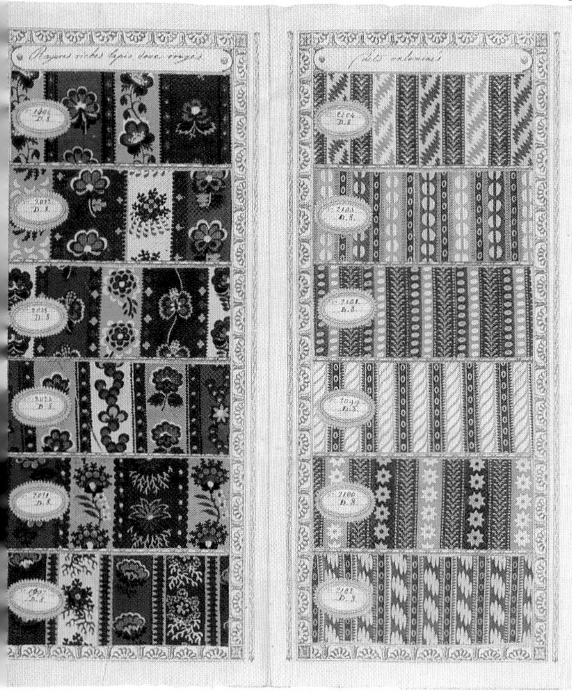

5

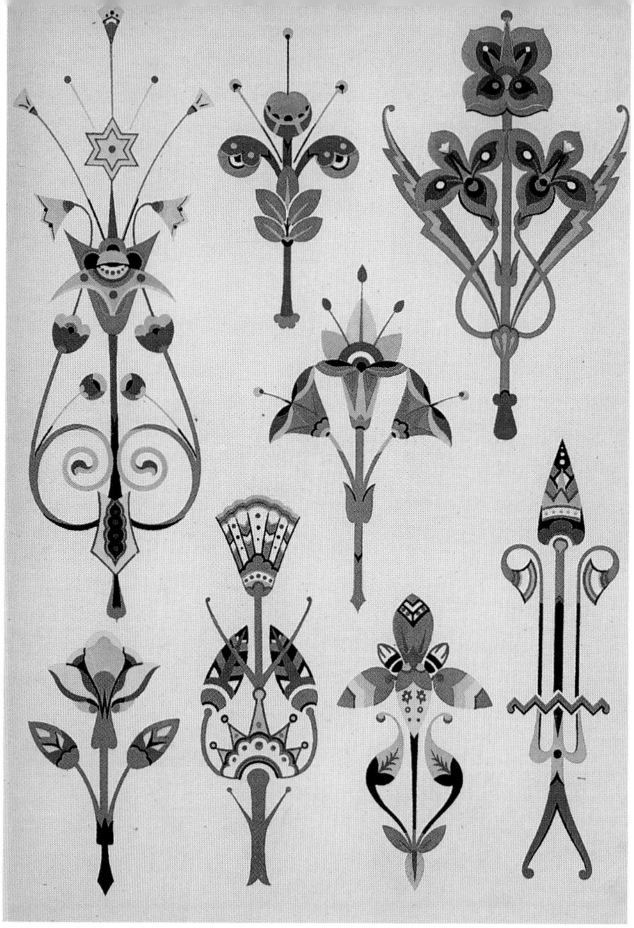

6

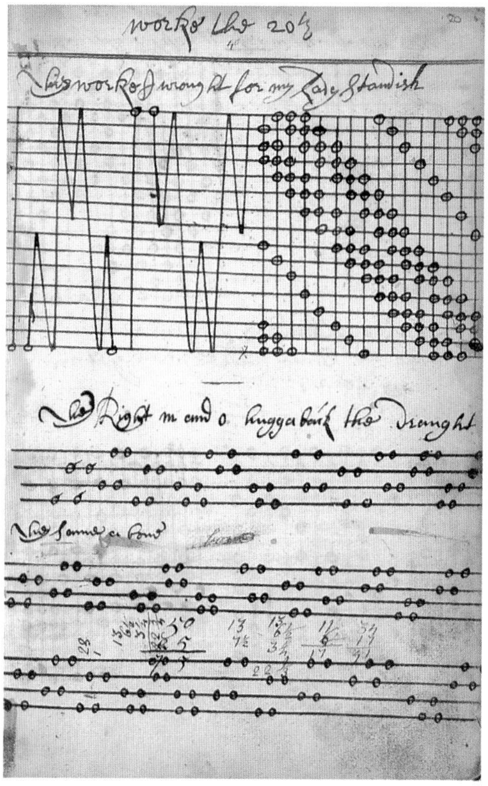

7

8

5 Tradesman's sample card of 19th-century French cotton dress fabrics. 6 Illustration from Christopher Dresser's *Studies in Design*, 1876. 7 Members of the Jackson family compiled this working reference book in England, from the late 17th through the late 18th century. They recorded the patterns they wove in a manner not unlike the notation of music. 8 An unfinished 17th-century English embroidery.

1

2

3

1 An inexpensive souvenir from the 1873 Vienna World Exposition, this fan shows the image of the Industrial Palace exhibition rotunda.
2 Jacques Charles launched an unmanned balloon from the Champs-de-Mars in Paris that landed several miles away near the town of Gonesse. This French furnishing fabric from 1784, entitled *"La Balon de Gonesse,"* records how terrified villagers attacked the "monster from the sky" with pitchforks, and the local curé was called to perform an exorcism. The fabric also records a less eventful flight from Paris to Nesle undertaken by the same Jacques Charles and a companion, the first to carry passengers. 3 The early success of the Soviet space program is commemorated by this small porcelain sculpture from the 1960s of two cosmonauts and a rocket.

Events and celebrations find their way into souvenirs of all kinds, to be remembered long after the fanfare has died down.

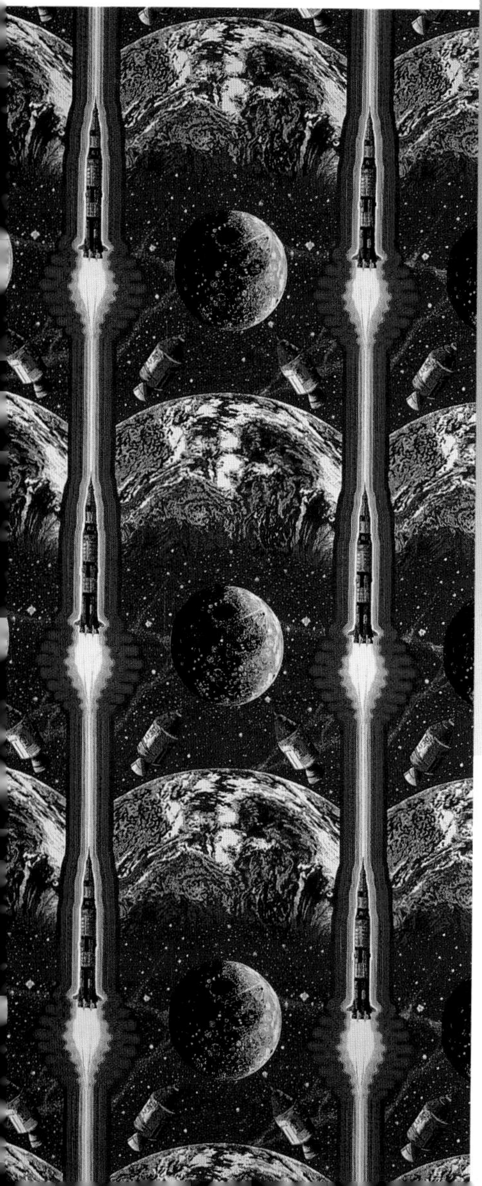

4

5

4 The 1969 Apollo 11 moon walk inspired Eddie Squires, then design director of Warner Fabrics in London, to produce this cotton, screen-printed fabric "Lunar Rocket." 5 English graphic designer Eric Ravilious created this Boat Race Day vase with scenes of the Thames River event for Wedgwood in 1938.

Boardgames, buttons, and books are enlisted in the war of ideas—propaganda.

2

1

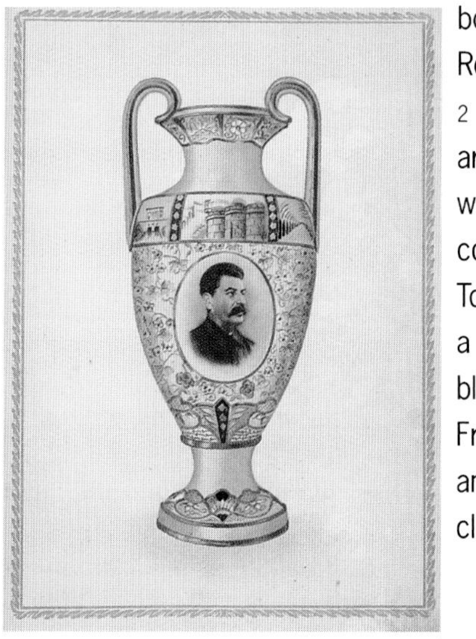

3

1 A 1922 Soviet propaganda chess set known as "The Reds and the Whites" or "The Communists and the Capitalists." The ceramic pieces include a White queen spilling a cornucopia of coins, a White king as Death, a Red king as a factory worker, a Red bishop in a Red Army uniform, a White pawn as a worker bound by chains, and a Red pawn as a harvester.
2 These buttons, c. 1796, are from a set of 18 that were worn on the dress coat of Haitian hero Toussaint de L'Ouverture, a former slave who led a black rebellion against French and British forces and became the self-proclaimed emperor of the island. The detailed scenes painted on the buttons depict Haitians conversing, working, and playing.
3 This 1940 vase design with a portrait of Stalin is from a state porcelain factory catalogue from the Ukraine.

1

2

1 This French Revolutionary wallpaper, c. 1792, features the Phrygian cap, cockades, Roman fasces, and oak wreaths that replaced symbols of monarchy. It is one of the earliest examples of a wallpaper used in service of the state. 2 This plate, with decoration by Sergei Chekhonin, was issued by the State Porcelain Factory in Petrograd in 1919, just after the Russian Revolution. Plates with propagandistic decoration were exhibited in shop windows, in Soviet government offices, and at international fairs.

3

3 Josephine de Beauharnais, the wife of Napoleon Bonaparte, was the recipient of this unmounted silk fan leaf in 1797, a gift from the City of Paris. It shows Napoleon being crowned by Victory and Abundance.

Emblems of state herald regimes extinct and extant.

4 The abdication of Edward VIII in 1936 meant that souvenirs such as this handkerchief had to be quickly redesigned with George VI's image and name instead. This is a rare version with Edward's portrait, made for a coronation that did not take place. 5 An oversized pitcher, made to commemorate the 1855 International Exposition in Paris, displays the coats-of-arms of participating countries.

4

5

POWER
THE
NERVE CENTRE
OF LONDON'S
UNDERGROUND
EMcKnight-Kauffer 1930.

1

2

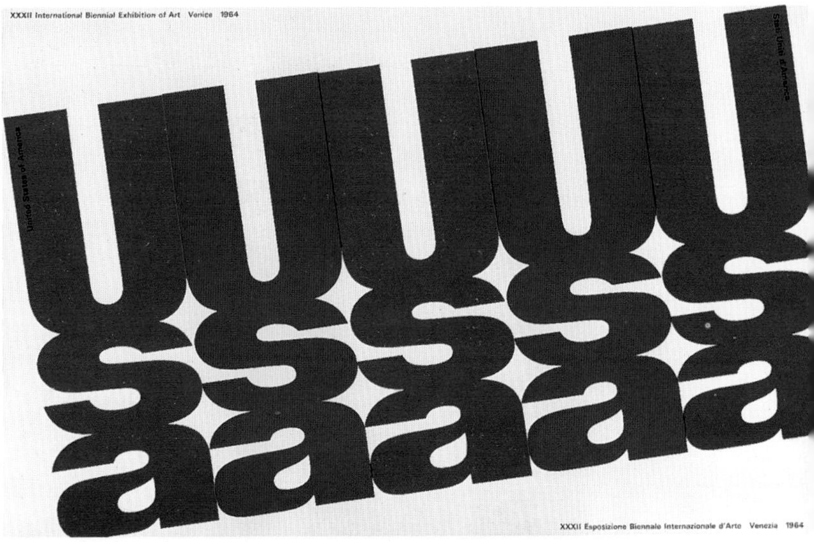

3

4

1 This 1930 poster, by Edward McKnight Kauffer, promotes the idea of harnessing energy for public transit with militant urgency. *Above:* 2 In this poster from 1937-41, Lester Beall contrasts the crisp modernity of a patriotic red, white, and blue pattern with the gritty realism of a black and white documentary photograph to underscore the Rural Electrification Administration's message that electricity will make rural life more wholesome, up-to-date, and efficient. 3 Graphic designer Elaine Lustig Cohen transformed the letters "USA" into an abstract design for the cover of the catalogue of American work shown at the XXXII Biennial International Exposition of Art in Venice in 1964. 4 A Cuban film poster designed by Luis Vega in 1972 advertises the Soviet film *El Dominio del Fuego* (The Domain of Fire).

The power to move, the potential to motivate: the poster compels with a force that is larger than its size.

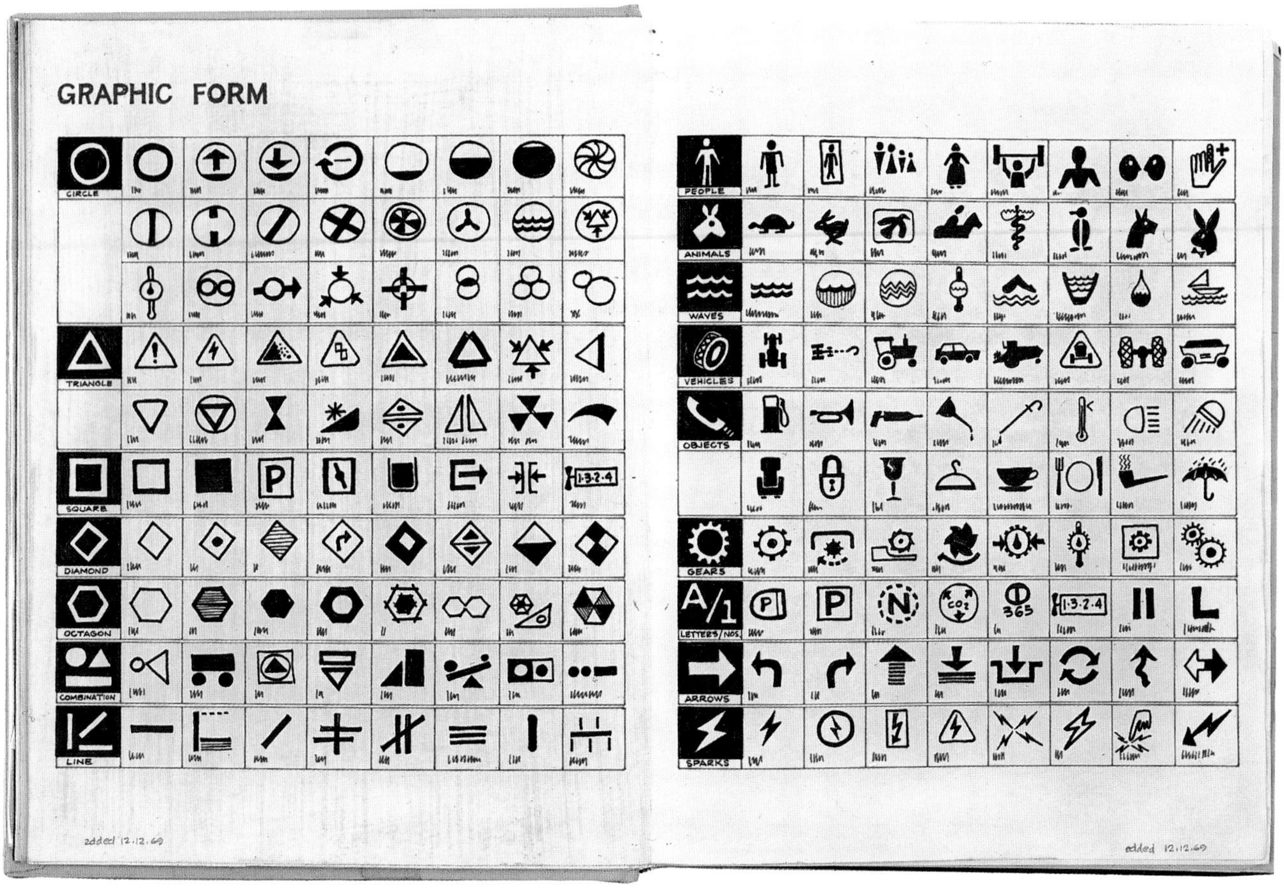

1

2

3

1 These pages on "Graphic Form" from Henry Dreyfuss's book *Symbol Sourcebook: An Authoritative Guide to International Graphic Symbols* illustrate his efforts to create a universal signage system. Dreyfuss's interest in pictographs originated in his industrial design work for companies that sold their products internationally and sought to transcend language barriers through symbols. Inspired by this marketing trend and his own life-long interest in visual communication, Dreyfuss conducted extensive research on symbolic language and published his standardized pictograms in 1972.

2 A version of Lumium Ltd. letterhead design, Edward McKnight Kauffer, 1935.

3 American Airlines *Fashions in Flight* poster designed by Edward McKnight Kauffer, c. 1947.

4

We communicate and navigate with a code of logos, symbols, emblems, and signs.

4 William Metzig designed the Monowatt logo in the 1940s for Westinghouse. He had a particular interest in heraldry and designed logos for a variety of corporations during his career as a graphic designer in Germany and, after 1939, the United States. 5 Best known for his Art Deco silver designs, Jean Puiforcat produced over 1,000 designs for monograms between 1930 and 1945, six of which are shown here. These stylish monograms are based on the geometry of triangles, squares, and circles.

5

1

2

1 Cheer box design, Donald Deskey Associates, c. 1957. 2 Matchsafes, designed to carry friction matches, were popular from the 1850s until about 1920. They were often used as commemoratives and for commercial advertising. 3 In the 1820s George Putnam and Amos Roff advertised their business on this bandbox, which was carried as a shopping bag would be today.

3

Colorful messages compete for the attention of consumers.

4

4 Fiorucci shopping bag, c. 1977. 5 Promotional fabric for the English company Bassett's Candies that was distributed in the 1930s and '40s to shops that sold the candy for use in window and case displays.

5

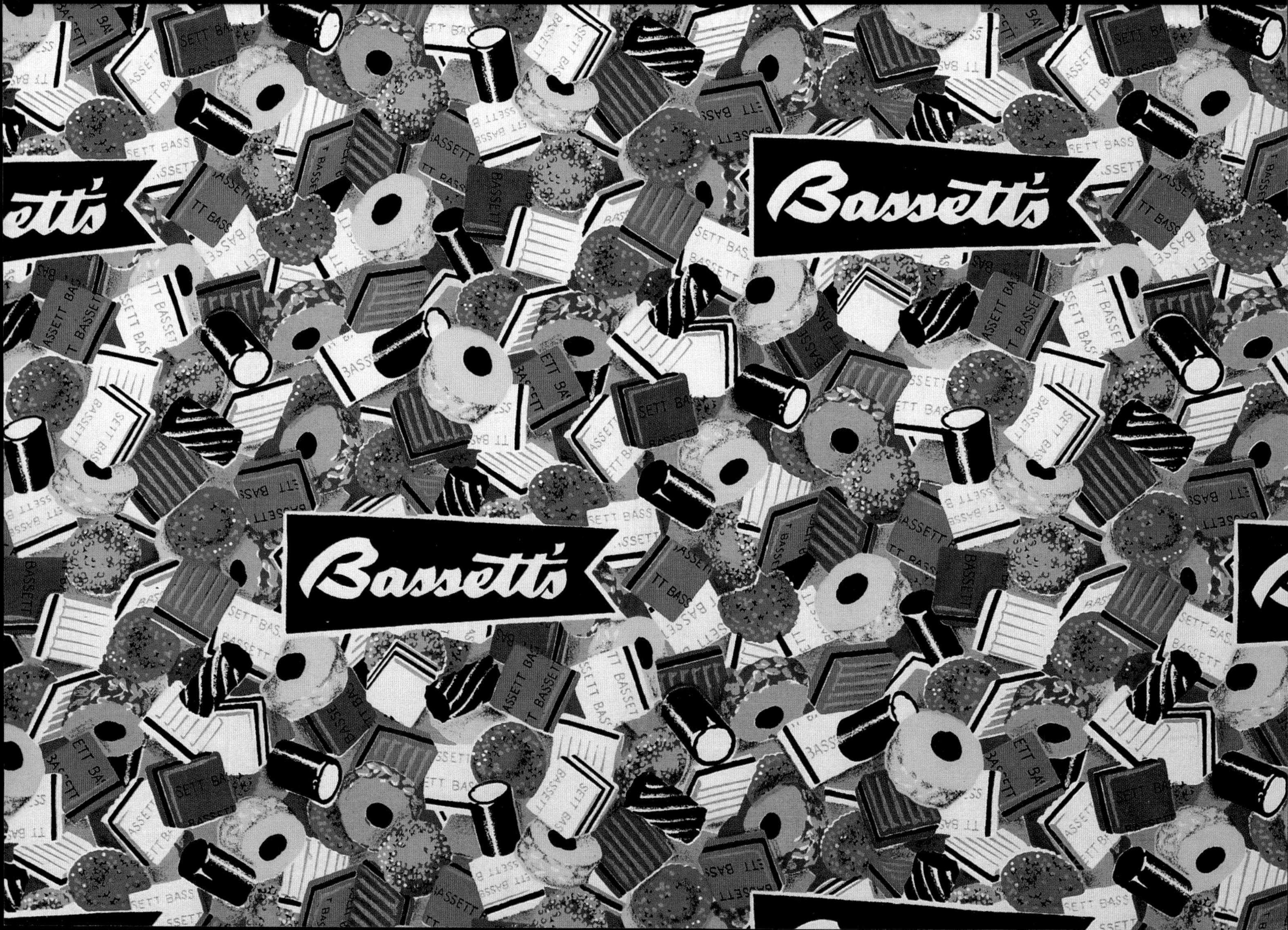

1

2

1 A souvenir from the 1939 New York World's Fair, this peep show gives the viewer an extended image of one of the fair's boulevards, complete with waterways and tree-lined walks toward the central Trylon and Perisphere. 2 Jayme Odgers and April Greiman's 1978 poster for the California Institute of the Arts is emblematic of the "New Wave" of graphic design they pioneered in the mid-1970s. Images and letterforms float through surreal spaces, mixing the conventions of design with an intensely personal vision. 3 *Death of a Salesman* poster, Paul and Carolyn Montie, 1993. 4 *The Edge of the Millennium* symposium poster for Cooper-Hewitt, National Design Museum, designed by Lorraine Wild, 1991.

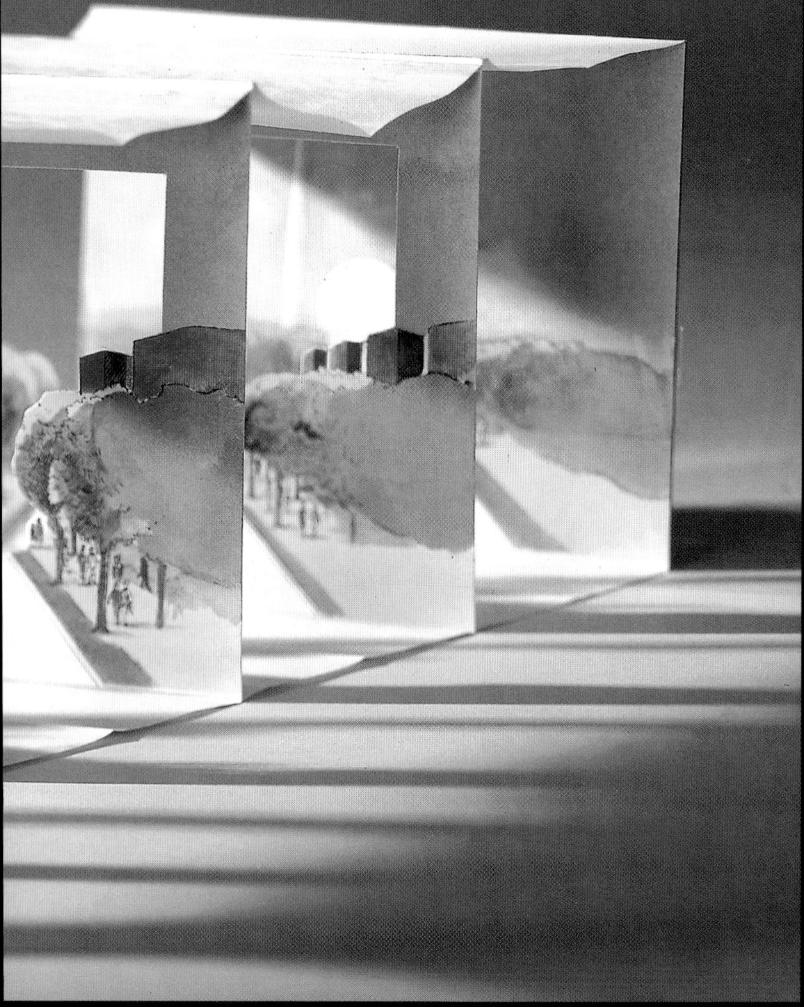

3

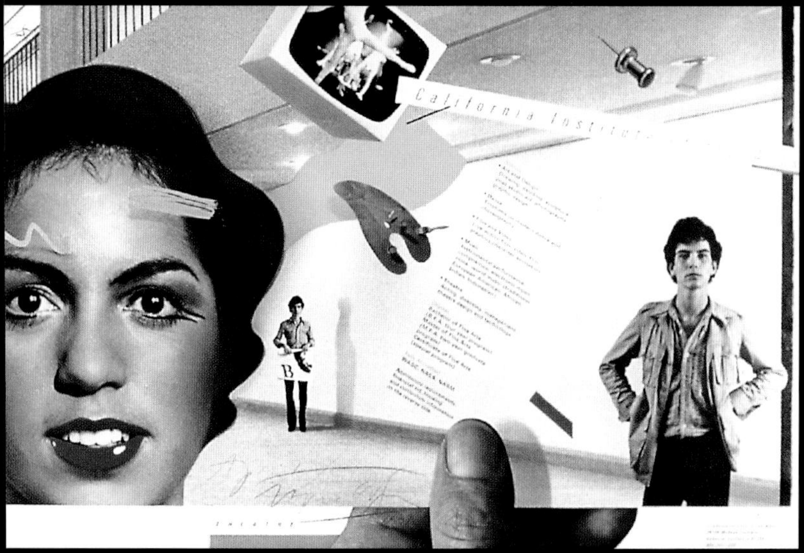

We warp space and
defy gravity by illusion.

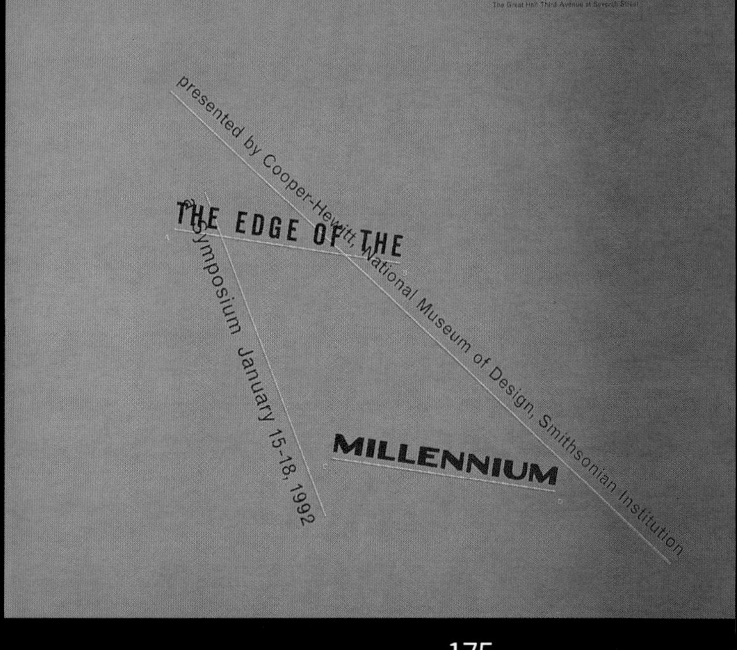

4

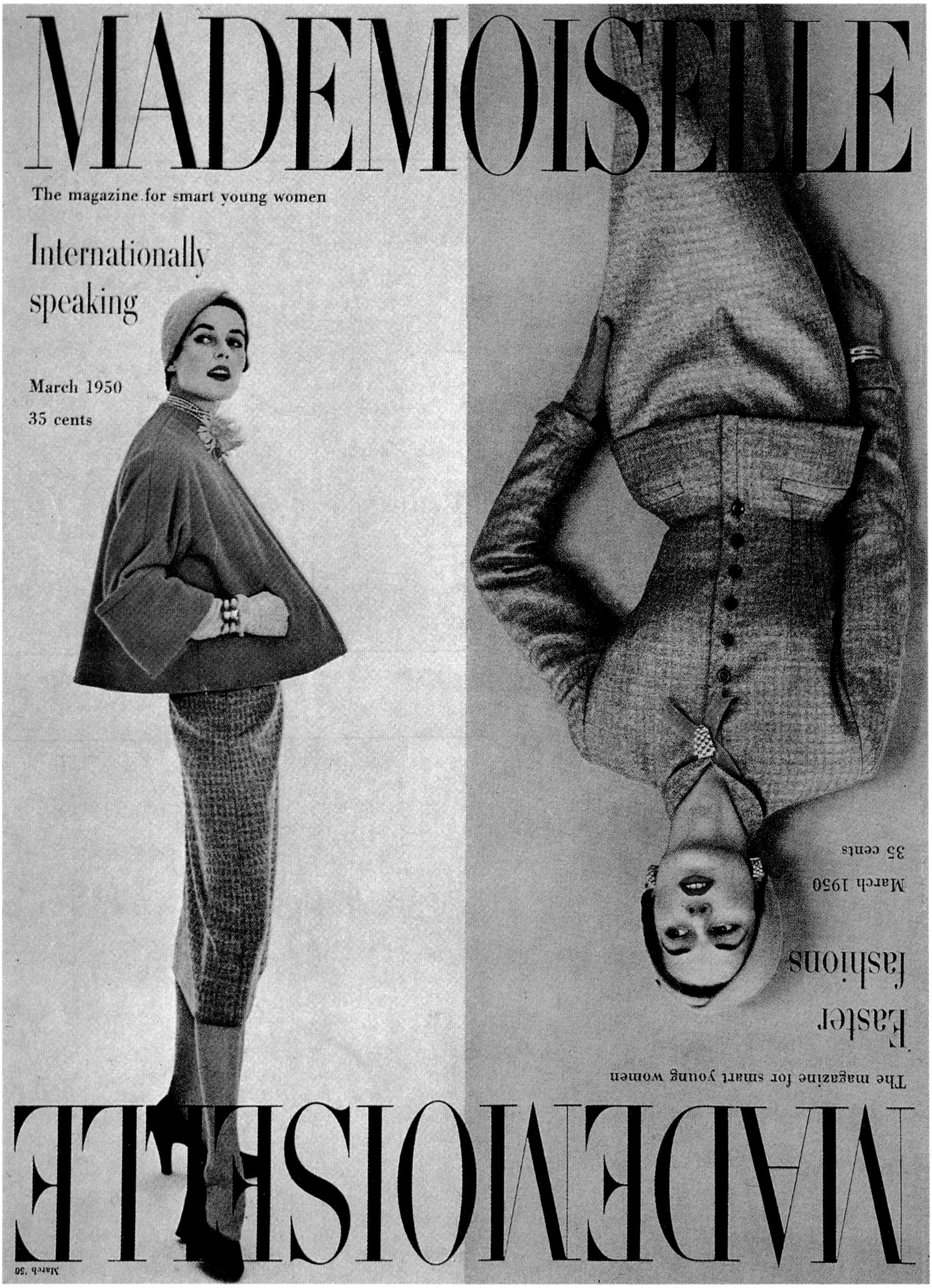
MADEMOISELLE
The magazine for smart young women
Internationally speaking
March 1950
35 cents
Easter fashions
March 1950
35 cents
The magazine for smart young women
MADEMOISELLE
March '50

1

Magazines mirror the world with topsy-turvy acuity.

1 Bradbury Thompson had a prolific career as a graphic designer and served as art director at *Mademoiselle* from 1945 to 1949. Thompson's cover for the March 1950 issue of *Mademoiselle* shows his sophisticated manipulation of photography, type, and color. The unique positioning of two sets of text, titles, and photographs next to and upside down from each other created an unconventional design that, at the same time, increased the magazine's newsstand visibility.

2 This amusing drawing by Christina Malman, proposed as a cover for either *The New Yorker* or *Promenade Magazine*,

1

2

1

1 Book cover of an 1867 edition of *Three Hundred Aesop's Fables.* 2 An enameled box in the form of a dalmation's head, England, c. 1850. 3 Metal toy grasshopper, c. 1900. 4 Originally published in 1806 as ornithological

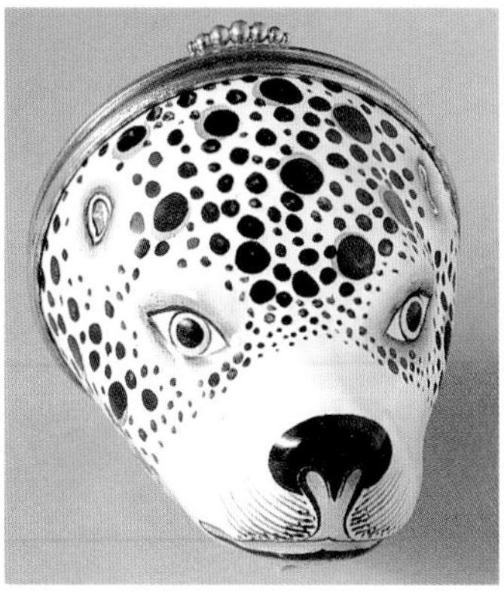
2

encyclopedias, François Le Vaillant's illustrated series on exotic birds, including this "Toucan with a Yellow Collar," became a vital source of inspiration for 19th-century textile and wallcovering designers.

3

Wildlife stories preserved in leather, porcelain, tin, and paper.

Le Tocan à Collier jaune. N.º 4.

Barraband pinx.t De l'Imprimerie de Langlois. Pérée sculp.t

Yellow Collar Toucan

4

Books are cabinets of curiosities awaiting exploration.

The Renaissance practice of collecting man-made and natural objects for display as curiosities continued into the 18th century. Albert Seba, a Dutch apothecary and amateur naturalist, assembled a large collection of natural specimens that was well known in his day. Between 1734 and 1765, his extensive natural menagerie was published in a lavish four-volume set. The pages shown here are indicative of the clever arrangement of images that can be found throughout the book.

Illustration and Photo Credits

Notes include title or name of work, designer, manufacturer, country, date, medium, donor or purchase credit, acquisition number, and photo credit. The first four digits of the acquisition number refer to the year of gift or purchase.

Abbreviations of Funding Sources:

GAE
General Acquisitions Endowment

SRCAPF
Smithsonian Regents Collections Acquisition Program Fund

DAAAF
Decorative Arts Association Acquisition Fund

Abbreviations of Photographers:

DC Dennis Cowley
MF Mike Fischer
MTF Matt T. Flynn
AG Andrew Garn
SH Scott Hyde
HI Hiro Ihara
DK Dave King
JP John Parnel
KP Ken Pelka
MR Michael Radtke
TR Tom Rose
VS Victor Schrager
ST Steve Tague
JW John White
CW Carmel Wilson

Design for Daily Life

Page 18

ATOMIC CLOCK. George Nelson Associates. Mfr'd by the Howard Miller Clock Company, Zeeland, MI. USA, 1949. Metal, wood. Gift of Mel Byars, 1991-26-1. Photo VS.

Pages 20-21

1. VASE. Grueby Pottery, Boston, MA. USA, c. 1905. Glazed earthenware. Gift of Marcia and William Goodman, 1983-88-7. Photo CW.
2. SUNBEAM MIXMASTER. Mfr'd by the Sunbeam Corporation. USA, c. 1950. Metal, glass, plastic. DAAAF, 1993-150-1. Photo DK.
3. SILVER FORK. George Washington Maher (American, 1864-1926). Mfr'd by Gorham Manufacturing Company, Providence, RI, for Spaulding & Co., Chicago, IL. USA, 1912. Silver, gilding. DAAAF and SRCAPF, 1995-49-7.

Pages 22-23

1. GOBLET. Tiffany & Company, NYC. USA, 1907. Gilt bronze. GAE, 1985-76-1. Photo DC.
2. COCKTAIL SHAKER. Emil A. Schuelke. Mfr'd by Napier. USA, 1936. Silver-plated metal. Gift of Rodman A. Heeren, 1971-92-1. Photo DK.
3. BEAKER. Switzerland, 18th century. Glass with enamel decoration. Gift of C. Helme Strater, Jr., John B. Strater, and Margaret S. Strater, 1976-1-31. Photo DC.
4. HAND CHATELAINE. USA, c. 1890. Gilded metal, enamel, glass, ivory. Gift of Mrs. Owen E. Robinson and Mrs. John B. Hendry, 1993-68-46. Photo DC.
5. BRACELET. England or USA, c. 1837. Gold, hair, pearls, glass. Gift of Mrs. Charles W. Lester, 1960-17-1. Photo JP.

Pages 24-25

1. ILLUMINATION SHOWING MARY IN GLORY. From a prayer book. The Netherlands, c. 1500. Photo MTF.
2. TORAH POINTER or *YAD*. Italy, 18th century. Silver. Gift of Ruth Friedman in memory of Harry G. Friedman, 1966-3-14. Photo HI.
3. LUGGAGE LABEL. 20th century. Letterpress on paper. Sarah Cooper Hewitt Fund, 1994-62-26. Photo MTF.
4. ILLUSTRATION FROM *CHAIR REQUIREMENTS FOR ELECTRA 188*, a report to Lockheed Corporation. Henry Dreyfuss (American, 1904-1972). USA, 1955. Henry Dreyfuss Collection. Gift of Doris and Henry Dreyfuss, 1972. Photo DC.

Pages 26-27

1. PUBLICITY PHOTOGRAPH OF DINING CAR APPOINTMENTS FOR THE 20TH CENTURY LIMITED TRAIN. Henry Dreyfuss (American, 1904-1972). USA, 1938. Henry Dreyfuss Collection. Gift of Doris and Henry Dreyfuss, 1972.
2. WALLPAPER WITH GIRL IN BATHING SUIT. Mfr'd by Marburger Tapeten. Germany, 1958. Machine-printed paper. Gift of Marburger Tapetenfabrik, 1958-96-2c. Photo MTF.

Pages 28-29

1. BADGE OF RANK. China, before 1644 (Ming Dynasty). Woven silk and gold threads; tapestry. Gift of J. P. Morgan (from the Miquel y Badia collection), 1902-1-432. Photo MTF.
2. MAN'S HAT. France or Italy, 18th century. Woven silk and metallic threads. Gift of Richard C. Greenleaf, 1952-47-2. Photo MTF.
3. FOLDING PLEATED FAN. England or France, 1800-10. Painted paper, parchment, ivory, mother-of-pearl, silver, metallic spangles. Anonymous gift, 1952-161-240. Photo MTF.

Pages 30-31

1, 2, 3. *LADIES' OLD FASHIONED SHOES*. Plates V, VIII, and IX. T. Watson Greig. Edinburgh: David Douglas, 1885. Photos DC, VS.

Pages 32-33

1. PLATE FROM *THE CYCLOPAEDIA OF THE BRITISH COSTUMES FROM THE METROPOLITAN REPOSITORY OF FASHIONS*. No. 1, Vol. 5. London: G. Walker, 1843. Photo MTF.
2. THE SKIRT OF A MAN'S CEREMONIAL DANCE COSTUME. Ngende or Ngongo subtribe, Kuba Kingdom. Zaire, Africa, 20th century. Resist-dyed raffia. Textile Department Fund, 1990-133-1. Photo MTF.

Pages 34-35

1. CANE WITH FAMOUS ACTRESSES. Fred Brown. Probably USA, 20th century. Carved and polychromed wood, silver, steel. Gift of John B. Scholz in memory of Walter R. Scholz, 1987-97-12. Photo DC.
2. MEN'S NECKWEAR AND SHIRTS. From *A.A. Vantine & Co., Inc. Spring and Summer Catalogue*, p. 20. NY: A.A. Vantine, 1918. Photo MTF.
3. From left to right: GENTLEMAN'S WAISTCOAT. France, c. 1790. Embroidered silk. Bequest of Richard C. Greenleaf, 1962-54-35. Photo ST.
GENTLEMAN'S WAISTCOAT. France, 1750-1760. Printed and painted silk. Bequest of Richard C. Greenleaf, 1962-54-11. Photo ST.
GENTLEMAN'S WAISTCOAT. France, 1770-1780. Embroidered silk. Bequest of Marian Hague, 1971-50-123. Photo ST.

Pages 36-37

1. "DIOMEDES" TEXTILE DESIGN. Dagobert Peche (Austrian, 1887-1923) for the Wiener Werkstätte. Austria, 1919. Watercolor, gouache on paper. SRCAPF, 1988-62-620, 621. Photo MTF.
2. DESIGN FOR A DRESS. From *Mode Wien*. Dagobert Peche (Austrian, 1887-1923) for the Wiener Werkstätte. Austria, 1914-15. Hand-colored woodcut on paper. Purchase, 1979-93-1. Photo MTF.
3. "BIG CHECKERBOARD" FABRIC. Junichi Arai (Japanese, b. 1932). Produced by Junichi Arai and Nuno Company, Tokyo. Japan, 1984. Woven nylon. Gift of K. K. Arai Creation System, 1990-63-4. Photo KP.
4. AINU ROBE. Made by a member of the Ainu people. Hokkaido, Japan, late 19th-early 20th century. Embroidered cotton with appliqué. Roy and Niuta Titus Foundation Fund and SRCAPF, 1994-92-1. Photo KP.

Pages 38-39

1. MAN'S LACE COLLAR. Italy, 1675-1700. Linen needle lace. Bequest of Richard C. Greenleaf, 1962-50-28. Photo VS.
2. EMBROIDERED MAN'S HAT. England, c. 1600. Silk and metallic threads on linen foundation. Bequest of Richard C. Greenleaf, 1962-53-11. Photo VS.
3. ALUMINUM JEWELRY. Probably France, c. 1855-60. Gold and aluminum. Gift of Mrs. Gustav E. Kissel, 1928-5-3a,b; Gift of Sarah Cooper Hewitt, Eleanor Garnier Hewitt, and Mrs. James O. Green in memory of their father and mother, Mr. and Mrs. Abram S. Hewitt, 1928-5-4a,b. Photo VS.

Pages 40-41

1. DRAW-STRING PURSE. England, early 17th century. Metallic yarns; macramé. Au Panier Fleuri Fund, Ida McNeil and GAE, 1989-30-1. Photo DC.
2. NECKLACE. Germany, 1930s. Silver, moonstone, marcasite. Gift of Sally Israel in memory of Fredericka Steinback, 1996-35-1. Photo DK.
3. ELDER'S NECKLACE. Kenya, Kikuyu people, 20th century. Shells, leather, metal, glass beads. Anonymous gift, 1995-37-8. Photo DC.
4. BELT BUCKLE. Attributed to Edward Colonna (German, 1862-1948, active France and USA). France, c. 1900. Retailed by Samuel Bing's shop, L'Art Nouveau, Paris, France. Gold, opal, pearls, garnets. Gift of Mr. and Mrs. Maxime Hermanos, 1967-88-3. Photo DC.
5. THREE SCENT CONTAINERS. Probably England or Germany, 18th century. Gold, enamel, agate, labradorite, shell cameos. Gift of the Panwy Foundation, Inc., from the collection of Maria Wyman, 1994-129-1,3,6. Photo HI.
6. HAIR ORNAMENT. Made by Tinfook. China, c. 1890. Gold, jade, pearls. Gift of Mrs. William Randolph Hearst, Jr., 1971-51-1. Photo DC.
7. HAND CHATELAINE. USA, c. 1890. Gilded metal, enamel, glass, ivory. Gift of Mrs. Owen E. Robinson and Mrs. John B. Hendry, 1993-68-46. Photo DC.
8. FOLDING COCKADE (CIRCULAR) FAN. "MANFG COMPANY LAMBERTVILLE GOODYEAR'S PATENT" inscribed on handle. USA, 1845-1860. Vulcanized rubber. Gift of Mrs. Henry Woodward Haynes, 1951-106-3. Photo MTF.
9. PURSE. Turkey, 19th century. Silk threads and wire; netting. Gift of Richard C. Greenleaf, 1951-105-19. Photo MTF.
10. "BASKET OF FLOWERS" BROOCH. England or France, c. 1800. Diamonds, silver, gold. Gift of Gertrude Sampson, 1962-26-1. Photo DC.

Pages 42-43

1. MICRO-MOSAIC PARURE AND BROOCH. Italy, c. 1810. Gold, hardstones. Gift of Frederick Saal in honor of Dr. and Mrs. Joseph Saal, 1991-160-1/8. Photo HI.
2. SALES BOOK OF SAMPLE BUTTONS. France, late 18th century. Gilded metal, metal foil, thread, and paillettes. Gift of Eleanor and Sarah Hewitt, 1921-22-72. Photo MTF.

3. HAND SCREEN. "Gio-Antonelli S. Apena" embossed on the screen. Venice, Italy, early 18th century. Engraved and embossed painted and gilded paper. Gift of the Museum Council, 1925-1-62. Photo DC.

Pages 44-45

1. THE "DOLPHIN" WATER CLOSET. From *Catalogue "G": Illustrating the Plumbing and Sanitary Department of the J. L. Mott Iron Works...*, pp. 72-73. NY: J. L. Mott Iron Works, 1888. Photo MTF.

2. TOILET CERAMICS. From *Grand Dépot Porcelaines, Faiences & Cristaux*, Plate 49. La Maison du Grand Dépot. Paris: Grand Dépot, 1889. Photo MTF.

3. PHOTOGRAPH OF A MANICURE AT THE ANTOINE BEAUTY SALON. Paris, c. 1929. Thérèse Bonney Collection, IBP009. Gift of Thérèse Bonney, 1939.

4. *THE CHILDREN'S BATH*. Israhel van Meckenem (German, 1440/1450-1503). Germany, before 1500. Engraving. Gift of Mrs. Leo Wallerstein, 1962-181-1. Photo MTF.

5. WOMAN'S HAIRSTYLE. From *L'Art de la Coëffure des dames Françoises*, by M Legros. Paris: Antoine Boudet, 1768, pp. 75-76. Photo MTF.

6. S50 STANDARD ELECTRIC SHAVER. Mfr'd by Braun AG. Germany, 1950. Plastic housing. Gift of Barry Friedman and Patricia Pastor, 1986-99-33a,b. Photo DK.

7. DISPOSABLE RAZORS. Mfr'd by Gillette and Schick. USA, 1980s. Plastic, metal. Anonymous gift, 1991-91-11, 19a. Photo DK.

8. WALLY RAZOR. Jack Hokanson (American, b. 1957). Hoke 2, Campbell, CA. USA, 1992. Rubber, metal. Gift of Hoke 2, 1993-72-1. Photo HI.

9. RADIUS TOOTHBRUSH. James O'Halloran (American, b. 1942). Mfr'd by Radius, Kutztown, PA. USA, 1983. Cellulose, nylon. Anonymous gift, 1991-91-25. Photo HI.

Pages 46-47

1. TEMPLE HANGING. Tibet, 18th or early 19th century. Silks, metallics, gilt paper; assemblage of Chinese fabrics. SRCAPF, 1987-7-1. Photo KP.

2. *VIEW OF THE PULPIT, SAINT SULPICE, PARIS.* Charles de Wailly (French, 1730-1798). France, 1789. Pen and ink, watercolor, chalk on paper. Gift of the Museum Council, 1911-28-293. Photo MTF.

Pages 48-49

1. TEMPLE CARVING. Thailand, possibly 18th century. Carved, polychromed, and gilded wood. Gift of Paul Manheim, 1967-92-9. Photo DC.

2. TORAH POINTER or *YAD*. Italy, 18th century. Silver. Gift of Ruth Friedman in memory of Harry G. Friedman, 1966-3-14. Photo HI.

3. LACE BORDER. Portugal, c. 1600. Linen, cut work, and needle lace. Bequest of Richard C. Greenleaf, 1962-50-23. Photo DC.

4. FABRIC. Spain, late 14th or early 15th century. Woven silk. Gift of J. P. Morgan (from the Miquel y Badia Collection), 1902-1-302.

5. DESIGN FOR WOVEN SILK AND EMBROIDERY. Italy, 14th century. Pen and ink on vellum. Friends of Textiles and gifts of various donors, 1993-119-1. Photo KP.

Pages 50-51

1. LEAF FROM "THE ORDER FOR THE ADMINISTRATION OF THE LORD'S SUPPER, OR HOLY COMMUNION." From *The Altar Book....* Protestant Episcopal Church in the U.S.A. Boston: D. B. Updike, The Merrymount Press, 1896. Designs by Bertram Grosvenor Goodhue (American, 1869-1924), Robert Anning Bell (1863-1933), and Charles Sherborn. Photo AG.

2. ECCLESIASTICAL FABRIC. Spain, 17th century. Woven silk. Gift of J. P. Morgan (from the Miquel y Badia Collection), 1902-1-845.

3. *RETURN FROM THE BAPTISM*. From *Le Mariage à la Ville* (*Marriage in the City*), Plate IV. Abraham Bosse (French, 1602-1676). France, 1633. Etching and engraving. SRCAPF and GAE, 1996. Photo MTF.

4. CENSER. Italy, probably 18th century. Bronze. Gift of Mrs. Russel C. Veit, 1951-66-16. Photo HI.

5. DESIGN FOR A CHALICE. From *Disegni Diversi*. Maximilian Joseph Limpach (b. Prague), after Giovanni Giardini (Italian, 1646-1722). Italy, 1714, reissued 1750. Etching and engraving. Gift of F. Burrall Hoffman, 1960-13-3. Photo MTF.

6. PRINTED FURNISHING FABRIC OR DECORATIVE PANEL. A. Nedby, aged 10 (American). USA, 1930s. Screen-printed cotton. Gift of the estate of Ella Ostrowsky, 1976-98-20. Photo MTF.

7. ALB (ECCLESIASTICAL GARMENT). Italy, Spain, or Portugal, c. 1600. Linen, embroidery, and bobbin lace. SRCAPF, with the gift of Mrs. John Innes Kane, 1995-44-1. Photo KP.

Pages 52-53

1. KYLIX. Greece, 6th century B.C. Pottery, black-figure decoration. Anonymous gift, 1921-1-1. Photo DC.

2. *THE OPEN TABLE FOR THE THREE UPPERMOST CLASSES OF GENTLEMEN OF LOWER AUSTRIA.* Christian Engelbrecht (German, 1672-1752) and Johann Andreas Pfeffel (German, 1674-1748), after Johann Cyriak Jackjofer (German, 1675-1731). From *Erbhuldigung So Dem...römischen Kayser... als Ertz-Hertzogen zu Oesterreich Josepho dem Ersten...*, Plate VII. Germany, 1705. Engraving on paper. Purchase in memory of Samuel C. Harriman, 1953-5-2. Photo MTF.

3. CANDELABRA. Claude Ballin the Younger (French, 1661-1754). France, 1739-40. Silver. Anonymous gift, 1949-131-1, 2. Photo VS.

Pages 54-55

1. PLATE FROM *SPOONS, FORKS, KNIVES, ETC.: MADE BY THE MERIDEN BRITANNIA CO.... CATALOGUE NO. 43.* Meriden, CT: The Meriden Britannia Company, 1847. Photo DC.

2. KUBUS STORAGE SYSTEM. Wilhelm Wagenfeld (German, 1900-1990). Mfr'd by Vereinigte Lausitzer Glaswerke. Germany, 1938. Glass. Sir Arthur Bryan Fund and GAE, 1990-4-1/18. Photo VS.

Pages 56-57

1. PLATE. Mfr'd by Copeland & Garrett, Spode, Staffordshire. England, c. 1830. Porcelain, enamel, gilding. Bequest of Emily H. Chauncey, 1959-155-1a. Photo DC.

2. "CANTON" PLATE. Edward Colonna (German, 1862-1948, active France and USA). Mfr'd by Gérard, Dufraissex and Abbot, Limoges. Retailed by Samuel Bing's shop, L'Art Nouveau, Paris. France, 1899-1900. Porcelain. GAE, 1989-5-2. Photo DC.

3. PLATE. Gabriel Pasadena (American). USA, 1950. Glazed ceramic. Gift of Mel Byars, 1990-15-9. Photo DC.

4. DESSERT PLATE. Mfr'd by C. H. Pillivuyt & Cie. France, c. 1875. Porcelain, enamel, gilding. DAAAF, 1993-124-4. Photo DC.

5. PLATE. Mfr'd by Lavenia, Laveno. Italy, 1925-30. Porcelain, enamel, gilding. Gift of Daniel Morris and Denis Gallion, 1993-134-21. Photo DC.

6. PLATE SHOWING THE SÈVRES PORCELAIN FACTORY. Made by the Sèvres Porcelain Factory. France, 1822. Porcelain, enamel, gilding. Gift of George B. and Georgiana L. McClellan, 1936-13-32. Photo DC.

7. PLATE. Mfr'd at Rörstrand. Sweden, mid-19th century. Earthenware with transfer-printed decoration. Gift of Paul F. Walter, 1993-64-2. Photo DC.

8. "TRAVEL" PLATE. Eric Ravilious (English, 1903-1942). Mfr'd by Wedgwood, Staffordshire. England, 1938. Porcelain, printed decoration. Gift of Paul F. Walter, 1992-42-3. Photo DC.

9. PLATE. Simon Lissim (Russian, 1900-1981, active France and USA). Decorated by L. Rodzianko. France, 1927. Porcelain, enamel, gilding. Bequest of Simon Lissim, 1981-37-2. Photo DC.

10. PLATE. Jutta Sika (Austrian, 1877-1964). Koloman Moser School. Mfr'd by Josef Böck Porcelain Factory, Vienna. Austria, 1901-02. Porcelain, enamel. The Charles E. Sampson Memorial Fund, 1986-83-5. Photo DC.

11. "HOMEMAKER" PLATE. Mfr'd by Ridgeway Potteries, Ltd., Staffordshire. England, c. 1955. Glazed earthenware with printed decoration. Anonymous gift, 1986-78-1. Photo DC.

12. PLATE. Richard Riemerschmidt (German, 1868-1957). Mfr'd by the Meissen Porcelain Factory. Germany, 1903-5. Porcelain. DAAAF, 1996. Photo DC.

13. "FIESTA" AND "HARLEQUIN" TABLEWARES. Frederick Hurten Rhead (American, 1880-1942). Mfr'd by the Homer Laughlin China Company. USA, 1940s-50s. Glazed ceramic. Gift of Paul F. Walter, 1991-68-1, 2, 6, 7, 8, 9; Gift of Mel Byars, 1991-59-169, 170. Photo DC.

14. PLATE. Jean Luce (French, 1895-1964). Decorated by Charles Ahrenfeldt (1859-1969). France, 1937. Porcelain, enamel, gilding, silvering. Gift of James M. Osborn, 1969-97-14a. Photo DC.

15. QUATREFOIL CENTERPIECE. Massimo Vignelli (Italian, b. 1931, active USA from 1965). Produced by Heller Designs, Inc. USA, c. 1970. Plastic. Gift of Lella and Massimo Vignelli, 1986-111-21. Photo DK.

16. PLATE. Mfr'd by Schramberg Pottery. Germany, c. 1928-30. Earthenware. Gift of Victor Wiener, 1996-18-97. Photo DC.

17. PLATE. Mfr'd by Kornilov Brothers, St. Petersburg. Russia, 1900-1910. Retailed by Tiffany & Company, NYC. Porcelain, enamel, gilding. Anonymous gift, 1956-188-1a. Photo DC.

18. *"LES PÂTISSERIES"* PLATE. Mfr'd by the Sèvres Porcelain Factory. France, 1831. Porcelain, enamel, gilding. Bequest of Erskine Hewitt, 1938-57-323. Photo DC.

19. PLATE. Mfr'd by The Greek A Factory, Delft. The Netherlands, 1686-1701. Glazed earthenware. Bequest of Walter Phelps Warren, 1986-61-42. Photo DC.

20. PLATE. Peter Behrens (German, 1868-1940). Mfr'd by Bauscher Brothers Porcelain Factory. Wieden, Germany, c. 1901. Porcelain. Gift of Daniel Morris and Denis Gallion, 1993-134-13. Photo DC.

Pages 58-59

1. DESIGN FOR "HIGHLIGHT/PUNCH" FLATWARE, ALSO KNOWN AS "AMERICAN MODERN." Russel Wright (American, 1904-1976). For John Hull Cutlers Corporation. USA, 1950. Gouache, chalk, and ink on paper. Gift of Russel Wright, 1976-15-9. Photo KP.

2. "MUSEUM" CUP AND SAUCER. Eva Zeisel (Hungarian, b. 1906, active USA from 1930s). Mfr'd by Castleton China. USA, 1946. Porcelain. Gift of Paul F. Walter, 1993-30-10ab. Photo DC.

Pages 60-61

1. GOOD GRIPS KITCHEN UTENSILS. Designed by Davin Stowell, Tucker Viemeister, Dan Formosa, Stephen Russak, Stephan Allendorf, Michael Calahan, Jürgen Laub, and Stephen Wahl of Smart Design, Inc., and Sam Farber of OXO. Mfr'd by OXO International. USA, 1991. Stainless steel, santoprene (thermoplastic rubber), ABS plastic. Gift of OXO, 1992-52-23, 24, 25, 28. Photo VS.

2. "EAT/DRINK" TABLEWARE. Ergonomi Design Gruppen. Mfr'd by RFSU Rehab. Sweden, 1980. Metal, plastic. Gift of RFSU Rehab, 1982-76-1/8. Photo VS.

3. STUDIES FOR "DESIGN ONE" FLATWARE. Don Wallance (American, 1909-1990). For H. E. Lauffer. USA, 1953. Don Wallance Collection.

Pages 62-63

1. SET OF SIX *CALICI NATALE* GOBLETS. Carlo Moretti (Italian, b. 1934). Mfr'd by Carlo Moretti Studio, Murano. Italy, 1990. Glass. Gift of Carlo Moretti, 1991-75-1/6. Photo VS.

2. DESIGN FOR A COVERED SOUP TUREEN. Louis Süe (French, 1875-1968) and André Mare (French, 1885-1932). For Compagnie des Arts Français. France, 1921. Graphite, color pencil on paper. Purchased through the gift of Mrs. George A. Hearn, 1992-70-3. Photo MTF.

3. GOBLET. Tiffany & Company, NYC. USA, 1907. Gilt bronze. GAE, 1985-76-1. Photo DC.

4. SILVER FLATWARE. George Washington Maher (American, 1864-1926). Mfr'd by Gorham Manufacturing Company, Providence, RI, for Spaulding & Co., Chicago, IL. USA, 1912. Silver, gilding. DAAAF and SRCAPF, 1995-49-2, 4, 7, 9, 12. Photo VS.

Pages 64-65

1. COCKTAIL SHAKER. Emil A. Schuelke. Mfr'd by Napier. USA, 1936. Silver-plated metal. Gift of Rodman A. Heeren, 1971-92-1. Photo DK.

2. COVER OF *THE SAVOY COCKTAIL BOOK*. H. Craddock. London: Constable & Company, Ltd., 1930. Photo VS.

3. PHOTOGRAPH OF BARTENDER AT LE GRAND ECART. Paris, c. 1925. Thérèse Bonney Collection. Gift of Thérèse Bonney, 1939.

Pages 66-67

1. SHERRY GLASS. Mfr'd by Lobmeyr Glassworks. Austria, c. 1925. Blown glass. Gift of Denis Gallion and Daniel Morris, 1993-134-5. Photo VS.

2. GOBLET. Ireland, early 19th century. Blown and cut glass. Bequest of Mrs. John Innes Kane, 1926-22-497. Photo VS.

3. "EMBASSY" CHAMPAGNE GLASS. Walter Dorwin Teague (American, 1883-1960) and Edwin W. Fuerst. Mfr'd by Libbey Glass Company. USA, 1939. Molded, pressed, and etched glass. The Sir Arthur Bryan Fund, 1988-58-2. Photo VS.

4. "ARKIPELAGO" CORDIAL GLASS. Timo Sarpaneva (Finnish, b. 1926). Mfr'd by Iittala Glassworks. Finland, 1979. Blown and molded glass. Gift of Iittala, 1985-96-5. Photo VS.

5. "AMERICAN MODERN" GOBLET. Russel Wright (American, 1904-1976). Morgantown (West Virginia) Glass Guild. USA, introduced in 1951. Molded glass. Gift of Paul F. Walter, 1991-30-2. Photo VS.

6. GOBLET. Tiffany Studios, Corona, NY. USA, c. 1900. Favrile glass. Bequest of Joseph L. Morris, 1966-55-7c. Photo VS.

7. GOBLET. England, mid-18th century. Blown glass, opaque-twist stem. Gift of Mr. and Mrs. Thomas L. Wolf, 1979-62-1. Photo VS.

8. GOBLET. France or Belgium, c. 1900. Glass with etched decoration. Gift of Justin G. Schiller, 1992-100-15. Photo VS.

Pages 68-69

1. *CALENDRIER PERPÉTUEL*. France, 1790-1800. Hand-colored engraving and etching on paper. Purchase in memory of Mrs. John Innes Kane, 1945-18-3. Photo MTF.

2. KITCHEN TIMER. Mads Odgård (Danish, b. 1960) and Bernt Nobert (Danish). Mfr'd by Rosti Housewares. Denmark, 1992. Plastic housing. Gift of Marilyn F. Symmes, 1993-99-1. Photo DK.

3. DESIGN FOR CYCLOMETER CLOCK. George Nathan Horwitt (Romanian, 1889-1990, active USA from 1920s). USA, 1930s. Horwitt Collection. Photo MTF.

4. ZEPHYR CLOCK. Kem Weber (American, 1889-1963). Mfr'd by Lawson Time, Inc. USA, c. 1930. Copper, brass, plastic. DAAAF, 1994-73-3. Photo DC.

5. DESIGN FOR A CLOCK. Belgium, 1800-1825. Pen and ink, watercolor on paper. Gift of Eleanor and Sarah Hewitt, 1931-64-131. Photo KP.

6. TWO-TIME WRIST WATCH. Tibor Kalman (American, b. Hungary, 1949). Produced by M&Co. USA, 1980s. Stainless steel, metal with matte black finish, glass, and leather. Gift of Tibor Kalman, M&Co., 1993-151-5. Photo AG.

Pages 70-71

1. *MY LADY'S CHAMBER*. Walter Crane (English, 1845-1915). Frontispiece from *The House Beautiful* by Clarence Cook. NY: Scribner, Armstrong and Company, 1878. Photo MTF.

2. DOSCO NEEDLE CATALOGUE. Trade publication. Photo VS.

3. DARNING SAMPLER. Worked by PP (signed with initials only). The Netherlands, dated 1735. Embroidered silk on cotton. Bequest of Gertrude M. Oppenheimer, 1981-28-234. Photo DC.

4. SILVER STREAK IRON. Mfr'd by Corning Glass Works and Saunders Machine & Tool Corporation. USA, 1940-45. Glass, metal. DAAAF, 1996. Photo VS.

5. *A BETTER HOME* POSTER. For Rural Electrification Administration, U.S. Department of Agriculture. Lester Beall (American, 1909-1969). USA, c. 1937-1941. Offset lithograph and screenprint. SRCAPF, GAE, Friends of Drawings and Prints, Sarah Cooper Hewitt Fund, and through the gift of Mrs. Edward C. Post, 1995-106-2. Photo KP.

Pages 72-73

1. MODEL #302 TELEPHONE. Henry Dreyfuss (American, 1904-1972). Mfr'd by Western Electric Co., for Bell Telephone. USA, 1937. Metal housing. DAAAF, 1994-73-2. Photo HI.

2. *HELLO—THE TELEPHONE AT YOUR SERVICE* POSTER. For the British General Post Office. Edward McKnight Kauffer (American, 1890-1954, active England from 1914). England, 1937. Gouache over graphite on paper. Gift of Mrs. E. McKnight Kauffer, 1963-39-523. Photo MTF.

3. PHOTOGRAPH OF MOD II PICTUREPHONE®. For Bell Telephone Laboratories. Henry Dreyfuss Associates. USA, 1967. Henry Dreyfuss Collection. Gift of Doris and Henry Dreyfuss, 1972.

4. HAMMOND MULTIPLEX TYPEWRITER. Mfr'd by Hammond. USA, 1919. Metal, wood. DAAAF, 1994-73-4. Photo VS.

5. TETRACTYS 24 ELECTRONIC CALCULATOR. Marcello Nizzoli (Italian, 1887-1969). Mfr'd by Olivetti. Italy, 1956. Plastic housing. Gift of Barry Friedman and Patricia Pastor, 1986-99-4. Photo DK.

6. VALENTINE PORTABLE TYPEWRITER. Ettore Sottsass, Jr. (Austrian, b. 1917, active Italy from 1947) with Perry King (English, b. 1938). Mfr'd by Olivetti. Italy, 1969. Plastic housing. Gift of Barry Friedman and Patricia Pastor, 1986-99-40a,b. Photo HI.

7. NeXT COMPUTER. frogdesign. Mfr'd by NeXT. USA, 1986. ABS plastic and magnesium housing. Gift of frogdesign Inc., 1995-36-1, 2. Photo HI.

Pages 74-75

1. *SEPULCHER IN EGYPTIAN STYLE, WITH CARYATIDS AND A LION; SEPULCHER IN EGYPTIAN STYLE WITH DEATH CARRYING A LAMP; SEPULCHER IN EGYPTIAN STYLE WITH SPHINXES AND AN OWL*. Louis-Jean Desprez (French, 1743-1804, active Sweden). France, c. 1779-1784. Pen and ink, wash, graphite on paper. GAE and Eleanor G. Hewitt Funds, 1938-88-3950, 3952, 3953. Photo KP.

2. QUEEN'S SERVANT COSTUME DESIGN. For *Pisanella, ou La Morte Parfumée*, by Gabriel d'Annunzio. Léon Bakst (Russian, 1866-1924, active Paris). France, 1913. Graphite, watercolor, and metallic paint on paper. Gift of Mrs. Gilbert Macculloch Miller, 1947-77-1.

3. SQUID COSTUME DESIGN. For *Sadko*, by Nikolai Andreyevich Rimsky-Korsakov. Serge Soudeikine (Russian, 1886-1946, active USA). USA, c. 1929. Gouache, graphite on paper. GAE, 1984-65-2. Photo DC.

4. COSTUME FOR A *BOYAR*. Alexandra Exter (Russian, 1882-1949). Russia, 1921. Gouache, graphite on paper. Bequest of Simon Lissim, 1981-37-18. Photo MTF.

5. THREE COMPETITION DESIGNS FOR LA FENICE THEATER, VENICE: LONGITUDINAL SECTIONS; TRANSVERSE SECTION. Italy, 1788. Pen and ink, wash, watercolor, graphite on paper. GAE and Eleanor G. Hewitt Fund, 1938-88-119, 120, 122. Photo KP.

Pages 76-77

1. BRIONVEGA LS502 PORTABLE RADIO. Richard Sapper (German, b. 1932, active Italy from 1958) and Marco Zanuso (Italy, b. 1916). Mfr'd by Brionvega. Italy, 1964. Plastic housing. Gift of Max Pine and Barbara Pine, 1994-59-3. Photo HI.

2. "JAZZ 3" WALLPAPER. Cuno Fischer. Mfr'd by Rasch Kunstler Tapeten. Germany, 1959-60. Machine-printed paper. Gift of Gebrüder Rasch & Co., 1960-116-1c. Photo KP.

3. *NEWPORT JAZZ FESTIVAL/NEW YORK* POSTER. Milton Glaser (American, b. 1929). USA, 1978. Gift of Milton Glaser, 1979-42-4. Photo DC.

4. CROSLEY D25BE CLOCK RADIO. Mfr'd by Crosley. USA, 1953. Plastic housing. Gift of Max Pine and Barbara Pine, 1993-133-33. Photo DC.

5. *GIVE PEACE A DANCE* POSTER. Art Chantry (American, b. 1954). USA, 1986. Offset lithograph on paper. Gift of Art Chantry, 1995-69-40. Photo MTF.

6. SC 7300 STEREO COMPONENT SYSTEM. Mfr'd by General Electric. USA, c. 1973. Wooden housing, metal, plastic, glass. Gift of Theresa Fitzgerald, 1995-154-1a/c. Photo DC.

7. DESIGN FOR A GENERAL ELECTRIC CLOCK RADIO. Richard Arbib (American, 1917-1995) with Donald Henry. USA, 1957. Pastel, watercolor, gouache, graphite on vellum. Purchased through the gift of Mrs. Griffith Bailey Coale, 1992-181-1.

Pages 78-79

1. BEAU BROWNIE CAMERA. Walter Dorwin Teague (American, 1883-1960). Mfr'd by Kodak. USA, c. 1930. Enameled metal and paper housing. Gift of Mr. and Mrs. Maurice Zubatkin, 1987-92-4a,b. Photo HI.

2. PLUS 126 CAMERA. USA and Austria, c. 1970. Plastic housing. Gift of Barry Friedman and Patricia Pastor, 1986-99-25. Photo DK.

3. ASSEMBLY PERSPECTIVE OF CAMERA. For Polaroid Corporation. Henry Dreyfuss (American, 1904-1972). USA, January 30, 1962. Crayon, pen and ink, pastel, ballpoint pen on tracing paper. Gift of Henry Dreyfuss, 1972-88-357. Photo MTF.

4. PURMA SPECIAL CAMERA. Raymond Loewy (French, 1893-1986, also active USA from 1919). Produced by Thomas De La Rue for Purma Cameras, Ltd. Lens manufactured by Beck. England, 1937. Bakelite housing. Gift of Barry Friedman and Patricia Pastor, 1986-99-24. Photo HI.

5. PUBLICITY PHOTOGRAPH FOR POLAROID. Henry Dreyfuss (American, 1904-1972). USA, 1963. Henry Dreyfuss Collection. Gift of Doris and Henry Dreyfuss, 1972.

Pages 80-81

1. PAPER DOLL. USA, c. 1850s. Hand-colored engraving on paper. Gift of Grace Lincoln Temple, 1937-22-13. Photo KP.

2. "WINNIE THE POOH" WALLPAPER FRIEZE. After drawings by Ernest H. Shepard (English, 1879-1976). Probably England, after 1926. Block-printed and air-brushed paper. Gift of Standard Coated Products, 1975-2-2.

3. PAPER MODEL FOR A PARK KIOSK. Pellerin et Cie. France, c. 1870. Hand-colored lithograph. Purchase in memory of Peter Cooper Hewitt, 1953-66-1.

4. *YELLOW SUBMARINE* FILM CEL. Heinz Edelmann (German, b. 1934) for United Artists Studio. USA, 1967-68. Acrylic paint, pen, and ink on plastic. Purchase in memory of Erskine Hewitt, 1969-73-1. Photo SH.

5. FURNISHING FABRIC. Tony Sarg (American, 1882-1942). USA, c. 1935. Screen-printed cotton. Gift of Margaret J. Gibson, 1936-40-1. Photo HI.

6. GAME BOARD. France, c. 1780. Painted paper. Gift of Eleanor and Sarah Hewitt, 1921-22-209. Photo HI.

7. "PINOCCHIO AND THE WHALE." From *The Pop-up Pinocchio*.... Harold Lentz, illustrations. Based on *Avventure di Pinocchio*, by Carlo Collodi. NY: Blue Ribbon, 1933, pp. 84-85. Photo VS.

Pages 82-83

1. A SETTEE. From *J.L. Mott Iron Works Illustrated Catalogue of statuary, fountains, vases, settees, etc.* NY: J. L. Mott, c. 1873. Photo DC.

2. WALLPAPER WITH GIRL IN BATHING SUIT. Mfr'd by Marburger Tapeten. Germany, 1958. Machine-printed paper. Gift of Marburger Tapetenfabrik, 1958-96-2c. Photo MTF.

3. *SUNLIGHT AND SHADOW*. Winslow Homer (American, 1836-1910). USA, 1872. Oil on canvas. Gift of Charles Savage Homer, Jr., 1917-14-7. Photo MF.

4. "EGYPTIAN GARDEN" FURNISHING FABRIC. Designed and stenciled by Lanette Scheeline (American, b. 1910). USA, c. 1939. Air-brushed stenciled cotton. Gift of Lanette Scheeline, 1984-56-1. Photo MTF.

5. PLATE. Tatiana Demorei. Mfr'd by Dmitrov Porcelain Factory. Soviet Union, 1937. Porcelain, enamel, gilding. The Ludmilla and Henry Shapiro Collection, partial gift and purchase, DAAAF and SRCAPF, 1989-41-30. Photo JP.

Pages 84-85

1. LUGGAGE LABELS. 20th century. Letterpress on paper. Sarah Cooper Hewitt Fund, 1994-62-11, 17, 26, 29, 30, 32, 33, 45, 47, 50, 57, 63. Photo MTF.

2. *AMERICAN AIRLINES: TO NEW YORK* POSTER. Edward McKnight Kauffer (American, 1890-1954, active England from 1914). USA, 1948. Color lithograph. Gift of Mrs. E. McKnight Kauffer, 1963-39-139. Photo MTF.

3. *THE INVISIBLE CITY* POSTER. For the International Design Conference, Aspen, Colorado. Ivan Chermayeff (American, b. 1932). USA, 1972. Gift of Ivan Chermayeff, 1981-29-59. Photo DC.

Pages 86-87

1. *GRAPHICS STANDARDS MANUAL*. For New York City Transit Authority. Unimark International, NYC. USA, 1970, pp. 12 and 64. Photo MTF.

2. PUBLICITY PHOTOGRAPH OF DINING CAR APPOINTMENTS FOR THE 20TH CENTURY LIMITED TRAIN. Henry Dreyfuss (American, 1904-1972). USA, 1938. Henry Dreyfuss Collection. Gift of Doris and Henry Dreyfuss, 1972.

3. ILLUSTRATIONS FROM *CHAIR REQUIREMENTS FOR ELECTRA 188*, a report to Lockheed Corporation. Henry Dreyfuss (American, 1904-1972). USA, 1955. Henry Dreyfuss Collection. Gift of Doris and Henry Dreyfuss, 1972. Photo DC.

4. DESIGN FOR A CONVAIR 880 AIRPLANE LAVATORY. For General Dynamics Corporation. Probably Glen Boyles for Dorothy Draper & Co., Inc. USA, 1957. Air-brushed gouache, paint, graphite on illustration board. Gift of Carleton Varney and Richard Hunnings, 1977-106-2. Photo MTF.

Pages 88-89

1. THE MARMON SIXTEEN SEDAN. From *The Designer's Story*, Vol. 1. Walter Dorwin Teague (American, 1883-1960). Indianapolis, IN: Marmon Motor Car Co., 1930. Photo MTF.

2. CONVAIR AUTOPLANE. Henry Dreyfuss (American, 1904-1972). USA, 1947. Henry Dreyfuss Collection. Gift of Doris and Henry Dreyfuss, 1972.

3. COVER OF *MAPS... AND HOW TO UNDERSTAND THEM*. 2nd edition prepared by Richard E. Harrison, J. Mca. Smiley, Henry Lent. NY: Consolidated Vultee Aircraft Corporation, 1943. Ladislav Sutnar Collection. Ladislav Sutnar Bequest, 1976. Photo MTF.

4. CLAY MODEL OF THE S-1 LOCOMOTIVE. For the Pennsylvania Railroad. Raymond Loewy (French, 1893-1986, also active USA from 1919). USA, 1936. Gift of the Pennsylvania Railroad through Samuel M. Vauclain, 1937-58-6.

5. DESIGN FOR THE S-1 LOCOMOTIVE. For the Pennsylvania Railroad. Raymond Loewy (French, 1893-1986, also active USA from 1919). USA, 1936. Graphite, pastel on tracing paper. Gift of the Pennsylvania Railroad through Samuel M. Vauclain, 1937-58-2.

Design for Shaping Space

Page 90

AXONOMETRIC VIEW OF INTERIOR. From *Farbige Räume und Bauten*, Plate 30. Wilhelm Jöker. Stuttgart: G. Siegle & Co., c. 1929. Photo MTF.

Pages 92-93

1. DESIGN FOR A GARDEN. Jan David Zocher the Younger (Dutch, 1791-1870). The Netherlands, 1856. Watercolor, gouache, pen and ink on paper. McNeil Acquisitions Fund, 1988-76-1. Photo MTF.

2. DESIGN FOR THE *DEMOCRACITY* EXHIBITION IN THE PERISPHERE. For the 1939 New York World's Fair. Henry Dreyfuss (American, 1904-1972). USA, 1939. Henry Dreyfuss Collection. Gift of Henry and Doris Dreyfuss, 1972-88-158,3. Photo MTF.

Pages 94-95

1. DESIGN FOR THE "TEN-DECK HOUSE." R. Buckminster Fuller (American, 1895-1983). USA, c. 1928. Mimeo (mimeograph) print hand-colored with watercolor. SRCAPF and GAE, 1991-53-1. Photo MTF.

2. DESIGN FOR "SPORTSHACK." Donald Deskey (American, 1894-1989). USA, 1940. Gouache, partially air-brushed, stenciled, ink and graphite on illustration board. Gift of Donald Deskey, 1988-101-1515.

3. *GENERAL VIEW OF BAYHAM*. From *Observations on the Theory and Practice of Landscape Gardening*.... Humphry Repton (English, 1752-1818). London: J. Taylor, 1803. Photo DC.

4. IMAGE OF A CEREMONY. From *Sixteen Japanese Ceremonies*. Kyoto: Kyoto Art Society, 1903. Photo MTF.

Pages 96-97

1. "THE RECONCILIATION OF VENUS AND PSYCHE." From "Psyche." Merry-Joseph Blondel (French, 1781-1853) and Louis Lafitte (French, 1770-1828). Originally produced in 1815-16 by Dufour et Compagnie, Paris; this edition printed by Desfossé et Karth, Paris. France, 1923. Gift of Mr. and Mrs. Abraham Adler, 1974-109-11a,b,c. Photo KP.

2. "ACANTHUS" WALLPAPER. William Morris (English, 1834-1896). Printed by Jeffrey & Co., Islington. England, 1875. Gift of Robert Friedel, 1941-74-20. Photo KP.

3. WALLPAPER WITH SPACE STATIONS AND ROCKETS. USA, c. 1950. Machine-printed paper. Gift of Suzanne Lipschutz, 1991-89-133. Photo KP.

Pages 98-99

1. *A COUPLE SEATED ON A BED*. Israhel van Meckenem (German, 1440/50-1503). Germany, c. 1495-1503. Engraving. Gift of Mrs. Leo Wallerstein, 1959-72-1. Photo MTF.

2. DESIGN FOR A STATE BEDCHAMBER. From *Second Livre d'Appartement*. Daniel Marot (French, 1663-1752, active the Netherlands). The Netherlands, c. 1702. Etching with engraving on paper. GAE, 1988-4-53. Photo MTF.

3. ONE OF A PAIR OF FIREDOGS. Attributed to Pierre Gouthière (French, 1732-1813/14). France, c. 1775. Gilt bronze, enamel. Gift of Mrs. Howard J. Sachs and Mr. Peter G. Sachs in memory of Miss Edith L. Sachs, 1978-168-69a. Photo TR.

4. "PH ARTICHOKE" HANGING LIGHT. Poul Henningsen (Danish, 1894-1967). Produced by Louis Poulson & Co., Copenhagen. Denmark, 1958. Copper, steel, enameled metal. Purchase, 1983-16-1. Photo DK.

5. ROCKING CHAIR. Charles Eames (American, 1907-1978) and Ray Eames (American, 1912-1988). Mfr'd by Herman Miller Furniture Company, Zeeland, MI. USA, c. 1950. Fiberglass, metal, wood, rubber. Gift of Barry Friedman and Patricia Pastor, 1986-99-46. Photo JW.

Pages 100-101

1. CHILD'S CHAIR. Charles Eames (American, 1907-1978) and Ray Eames (American, 1912-1988). Mfr'd by Herman Miller Furniture Company, Zeeland, MI. USA, 1944. Bent, laminated wood. Gift of Mrs. Eric Larrabee, 1991-144-1. Photo ST.

2. JEWEL CABINET. Mfr'd at the Sèvres Porcelain Factory. France, 1824-26. Porcelain with enameled and gilded decoration, gilt bronze, glass, wood. Bequest of the Reverend Alfred Duane Pell, by transfer from the National Museum of American History, Smithsonian Institution, 1991-31-2. Photo DC.

3. DESIGN FOR A MIRROR FRAME WITH MONOGRAM OF MARIE ANTOINETTE. Richard de Lalonde (French, active 1780-1796). France, c. 1787-1790. Pen and ink, wash over graphite on paper. Gift of the Museum Council, 1911-28-207. Photo MTF.

Pages 102-103

1. VIEW OF PARIS, c. 1654. From *Topographia Galliae....* Martin Zeiller (1589-1661). Frankfurt: Caspar Merian, 1655-1661. Photo DC.

2. VIEW OF THE CITY OF NUREMBERG. From *Liber chronicarum...*, Plate 100. Hartmann Schedel (German, 1440-1514). Woodcuts by Michael Wolgemut and Wilhelm Pleydenwurff. Nuremberg: Anthon Koberger, 1493. Photo HI.

3. PLAN OF CANTON, CHINA. From *Het gezantschap der Neêrlandtsche Oost-Indische Compagnie....* Johannes Nieuhof (Dutch, 1618-1672). Amsterdam: Jacob van Meurs, 1665. Photo DC.

4. MANHATTAN ZONING MAP. From *Plan for New York City, 1969: A Proposal. 4: Manhattan.* NY: New York City Planning Commission, 1969, p. 109. Photo MTF.

5. PARTIAL VIEW OF THE *DEMOCRACITY* EXHIBITION IN THE PERISPHERE. For the 1939 New York World's Fair. Henry Dreyfuss (American, 1904-1972). USA, 1939. Henry Dreyfuss Collection. Gift of Henry and Doris Dreyfuss, 1972-88-158, 2. Photo MTF.

Pages 104-105

1. *STUDY FOR THE MAXIMUM MASS PERMITTED BY THE 1916 NEW YORK ZONING LAW, STAGE 3.* Hugh Ferriss (American, 1889-1962). USA, 1922. Crayon, ink, wash over photostat, varnish. Gift of Mrs. Hugh Ferriss, 1969-137-3. Photo MTF.

2. THE HEIGHT OF THE SINGER BUILDING. From *Cities: The Forces that Shape Them.* Edited by Lisa Taylor. NY: Cooper-Hewitt Museum, 1982, p. 81.

3. THE COLONNADE CONDOMINIUMS, SINGAPORE: ISOMETRIC ELEVATION. Paul Rudolph (American, b. 1918). USA, 1980. Color pencil and graphite over diazo print. Purchase in memory of Erskine Hewitt, 1985-63-9. Photo KP.

4. DESIGN FOR THE "TEN-DECK HOUSE." R. Buckminster Fuller (American, 1895-1983). USA, c. 1928. Mimeo (mimeograph) print hand-colored with watercolor. SRCAPF and GAE, 1991-53-1. Photo MTF.

Pages 106-107

1. PERSPECTIVE DESIGN FOR A PAINTED CUPOLA OF A CHURCH. After Andrea Pozzo (Italian, 1642-1709). Italy, 1700-1725. Pen and ink, wash on paper. GAE and Eleanor G. Hewitt Fund, 1938-88-3461. Photo MTF.

2. DESIGN FOR A TEMPLE OF CURIOSITY. After Etienne-Louis Boullée (French, 1728-1799). Pen and ink, wash, graphite on paper. France, c. 1790. Gift of the Museum Council, 1911-28-463. Photo MTF.

3. AVIATION BUILDING BY WILLIAM LESCAZE AND J. GORDON CARR ASSOCIATES FOR THE 1939 NEW YORK WORLD'S FAIR. Hugh Ferriss (American, 1889-1962). USA, c. 1937. Chalk on paper. Gift of Mrs. Hugh Ferriss, 1964-5-7. Photo MTF.

4. DECORATIVE PANEL FOR JULY FESTIVAL ARCHITECTURE. After Félix Duban (1798-1870). France, 1835. Watercolor, gouache, pen, and graphite on paper. Purchased with assistance from Emily and Jerry Spiegel and Phyllis Dearborn Massar, 1991-17-14.

5. STAIRCASES, CHÂTEAU OF CHAMBORD. From *The Architecture of A. Palladio in Four Books...*, Book 1, Plate 42. Andrea Palladio (Italian, 1508-1580). London: John Watts, 1715. Gift of Abram S. Hewitt. Photo MTF.

Pages 108-109

1. COTTAGE NO. 13. From *Architectural Designs for Model Country Residences....* John Riddell. Philadelphia: Lindsay & Blakiston, 1861. Photo DC.

2. ELEVATION OF REAR FACADE, CASTEL D'ORGEVAL, PARC BEAUSÉJOUR, NEAR PARIS. Hector Guimard (French, 1867-1942). France, 1904. Pen, ink, and graphite on tracing paper. Gift of Mme Hector Guimard, 1950-66-5. Photo MTF.

3. "HOUSE FOR TWO GENERATIONS" PROJECT, SECOND VERSION. Arata Isozaki (Japanese, b. 1931). Japan, 1977. Screenprint on paper. Drawings and Prints General Fund, 1985-7-1. Photo MTF.

4. DESIGN FOR "SPORTSHACK." Donald Deskey (American, 1894-1989). USA, 1940. Gouache, partially air-brushed, stenciled, ink and graphite on illustration board. Gift of Donald Deskey, 1988-101-1515.

Pages 110-111

1. DOOR PLATES. Augustus Welby Northmore Pugin (English, 1812-1852). England, c. 1845. Brass. Purchase, 1975-72-1/3. Photo HI.

2. DOOR PLATE FOR THE GUARANTY BUILDING, BUFFALO, NY. Louis Sullivan (American, 1856-1924). USA, c. 1894. Iron. Gift of Roger Kennedy, 1979-77-9. Photo HI.

3. DESIGN FOR GATES OF THE PARIS MINT. Jacques-Denis Antoine (French, 1733-1801) and Workshop. France, c. 1776. Pen and ink, wash, and chalk on paper. Gift of the Marquis Val Verde de la Sierra, 1923-51-2.

4. PAIR OF GATES FOR THE CHANIN BUILDING, NYC. René Chambellan (French, active USA). USA, c. 1928. Iron, bronze. Gift of Marcy Chanin, 1993-135-1, 2. Photo HI.

5. LOCK. Jean Dutartre. Spain, c. 1700. Steel. Gift of the Museum Council, 1910-30-49a,b. Photo HI.

Pages 112-113

THREE IMAGES OF CEREMONIES. From *Sixteen Japanese Ceremonies.* Kyoto: Kyoto Art Society, 1903. Photo MTF.

Pages 114-115

1. *A COUPLE SEATED ON A BED.* Israhel van Meckenem (German, 1440/50-1503). Germany, c. 1495-1503. Engraving. Gift of Mrs. Leo Wallerstein, 1959-72-1. Photo MTF.

2. DESIGN FOR A STATE BEDCHAMBER. From *Second Livre d'Appartement.* Daniel Marot (French, 1663-1752, active the Netherlands). The Netherlands, c. 1702. Etching with engraving on paper. GAE, 1988-4-53. Photo MTF.

3. *CANOPY BED IN THE TURKISH STYLE WITH ALTERNATE SUGGESTION IN THE ITALIAN STYLE.* From *Recueil de differents meubles.* Matthew Liard (English, c. 1736-1782). Paris: Perruquier, after 1762. Etching. Gift of the Museum Council, 1921-6-84.

4. PLATES FROM *BEDSTEADS, COTS, ETC., IRON & BRASS IN FOUR-POST, TENT, CANOPY AND OTHER KINDS.* Export Catalogue No. 33. Birmingham, England: Fitter Brothers, c. 1906.

5. PHOTOGRAPH OF A BEDROOM EXHIBITED AT THE 1937 PARIS EXPOSITION. Maurice Barret (French). France, c. 1937. From *L'Art de Vivre....* NY: Cooper-Hewitt Museum, 1989, p. 137.

Pages 116-117

1. APARTMENT OF QUEEN ELIZABETH OF PRUSSIA, CHARLOTTENBURG PALACE, BERLIN. Elizabeth Pochhammer (German, active Berlin 1864-1880). Germany, 1864. Watercolor, gouache, paint, over graphite on paper. Purchase in memory of David Wolfe Bishop, 1957-98-2. Photo MTF.

2. DESIGN FOR A SALON WITH BLUE CHAIRS AND MAUVE CARPET. André-Léon Arbus (French, 1903-1969). France, 1930s. Gouache, paint, graphite on paper. Drawings and Prints Department James Amster Fund, 1989-1-1. Photo MTF.

3. *FARBIGE BLITZLICHTAUFNAHME DER ARBEITSNISCHE—OBERGESCHOSS.* From *Ein Wohnhaus.* Bruno Taut (German, 1880-1938). Stuttgart: W. Keller & Co., 1927. Photo MTF.

Pages 118-119

1. DESIGN FOR A MANTELPIECE. Pietro Camporese the Younger (Italian, 1792-1873). Italy, 1830. Pen and ink, watercolor, wash, graphite on paper. GAE and Eleanor G. Hewitt Funds, 1938-88-4300. Photo MTF.

2. T86 THERMOSTAT. Henry Dreyfuss (American, 1904-1972). Mfr'd by Honeywell. USA, introduced 1953. Metal, plastic. Gift of Honeywell, Inc., 1994-37-1. Photo HI.

3. ELECTRIC RADIATOR. René Coulon (French, b. 1908). Mfr'd by Saint-Gobain Glassworks. France, 1937. Tempered glass, chromed metal. The James Ford Fund, 1990-102-1. Photo JW.

Pages 120-121

1. OIL LAMP. Italy, 1st century B.C. Earthenware. Gift of the estate of David Wolfe Bishop, 1951-84-28. Photo DC.

2. DESIGN FOR TWO GILT BRONZE CHANDELIERS FOR THE PAVILION DE BAGATELLE, PARIS. François-Joseph Belanger (French, 1744-1818). France, 1777. Pen and ink, watercolor on paper. Gift of the Museum Council, 1921-6-62. Photo SH.

3. CANDLESTICK. Made by Richard Hood and Richard Hood, London. England, 1878-79. Silver. Bequest of Mrs. John Innes Kane, 1926-37-133. Photo DC.

4. "DRAGONFLY" TABLE LIGHT. Tiffany Studios, Corona, NY. USA, 1900-10. Leaded glass, gilt bronze. Gift of Margaret Carnegie Miller, 1977-111-1a,b. Photo DC.

5. HANGING LIGHT. Jac van den Bosch (Dutch, 1868-1948). Made for the shop, ët Binnenhuis, Amsterdam. The Netherlands, 1902. Iron, silk. DAAAF, 1994-67-13. Photo DC.

6. STANDING LIGHT. Edgar Brandt (French, 1880-1960). France, c. 1925. Bronze, alabaster. Gift of Stanley Siegel, from the Stanley Siegel Collection, 1975-32-2. Photo DC.
7. "FALKLAND" HANGING LIGHT. Bruno Munari (Italian, b. 1907). Mfr'd by Danese Milano. Italy, introduced 1964. Aluminum, elasticized fabric. Gift of Danese Milano, 1991-108-6a/c. Photo DC.
8. "PH ARTICHOKE" HANGING LIGHT. Poul Henningsen (Danish, 1894-1967). Produced by Louis Poulson & Co., Copenhagen. Denmark, 1958. Copper, steel, enameled metal. Purchase, 1983-16-1. Photo DK.
9. "CHROME BLENDER" TABLE LIGHT. Virginia Restemeyer (American, b. 1951). USA, 1989. Metal. DAAAF, 1992-1-1ab. Photo HI.
10. TABLE LIGHT. Mfr'd by CGM. France, 1980s. Plastic housing. Gift of Mel Byars, 1991-59-85. Photo DC.
11. THREE FLASHLIGHTS: "VIDOR," "GARRITY," "COMPACT INDUSTRIAL." Mfr'd by Rayovac, Garrity, Eveready. USA, 1980s. Plastic housings. Gift of Max Pine and Barbara Pine, 1993-133-7, 8, 21. Photo DC.

Pages 122-123

1. JEWEL CABINET. Mfr'd at the Sèvres Porcelain Factory. France, 1824-26. Porcelain with enameled and gilded decoration, gilt bronze, glass, wood. Bequest of the Reverend Alfred Duane Pell, by transfer from the National Museum of American History, Smithsonian Institution, 1991-31-2. Photo DC.
2. DESIGN FOR A SOFA IN ALCOVE WITH TENT CEILING, probably for the Prince of Wales' bedroom, Royal Pavilion, Brighton. Frederick Crace (English, 1779-1859). England, c. 1801-04. Pen and ink, watercolor on paper. Purchase in memory of Mrs. John Innes Kane, 1948-40-26. Photo MTF.
3. DESIGN FOR TWO ANDIRONS AND A SCONCE FOR THE PAVILION DE BAGATELLE, PARIS. François-Joseph Belanger (French, 1744-1818). France, 1777. Pen and ink, watercolor on paper. Gift of the Museum Council, 1921-6-61. Photo MTF.

Pages 124-125

1. DESIGN FOR THE WEST WALL OF THE MUSIC ROOM, ROYAL PAVILION, BRIGHTON. Frederick Crace (English, 1779-1859). England, 1817. Watercolor, gouache over graphite. Purchase in memory of Mrs. John Innes Kane, 1948-40-9a,b. Photo SH.
2. "ARABESQUE" WALLPAPER PANEL. Mfr'd by Jean-Baptiste Réveillon (France, 1725-1811). France, 1786. Block-printed paper. Purchase, 1994-101-12. Photo MTF.
3. FURNISHING FABRIC. France, 1810-1820. Woven silk. Purchase in memory of Ida McNeil, 1992-155-1. Photo HI.
4. ORNAMENTAL MOTIFS. From *Ornamente aller Klassischen Kunstepochen....* Wilhelm Zahn (1800-1871). Berlin: G. Reimer, 1843. Photo DC.

Pages 126-127

1. "THE RECONCILIATION OF VENUS AND PSYCHE." From "Psyche." Merry-Joseph Blondel (French, 1781-1853) and Louis Lafitte (French, 1770-1828). Originally produced in 1815-16 by Dufour et Compagnie, Paris; this edition printed by Desfossé et Karth, Paris. France, 1923. Gift of Mr. and Mrs. Abraham Adler, 1974-109-11a,b,c. Photo KP.
2. PAGE OF EGG AND DART BORDERS. From a wallpaper factory workbook. Mfr'd by Jean-Baptiste Réveillon (French, 1725-1811). Paris, c. 1785. Block-printed paper. GAE and DAAAF, 1986-34-1. Photo DC.

Pages 128-129

"EL DORADO" WALLPAPER. Eugène Ehrmann (1804-1896), George Zipélius (1808-1890), and Joseph Fuchs (1814-1888). Mfr'd by Zuber, Rixheim. France, 1848. Gift of Dr. and Mrs. William Collis, 1975-77-1/12. Photo KP.

Pages 130-131

1. *GENERAL VIEW OF BAYHAM.* From *Observations on the Theory and Practice of Landscape Gardening....* Humphry Repton (English, 1752-1818). London: J. Taylor, 1803. Photo DC.
2. "A LOVE GARDEN," GROUND PLAN FOR A HOUSE AND FORMAL GARDEN. Study for *Les Jardins,* 1914. Paul Vera (French, 1882-1954); draftsman, Eugène Verdeau. Paris: Emile-Paul Frères, 1919. Pen and brush and ink, watercolor, gouache, graphite. SRCAPF, 1991-58-14. Photo MTF.
3. FURNISHING FABRIC. Produced by Favre Petitpierre, Nantes. France, early 19th century. Printed cotton. Bequest of Elinor Merrell, 1994-80-14. Photo MTF.

Pages 132-133

1. *"DÉCOR CHASSE ET PECHE"* WALLPAPER. Wagner. Mfr'd by Lapeyre & Cie, Paris. France, 1839. Block-printed paper. Friends of the Museum, 1955-12-10. Photo KP.
2. "ACANTHUS" WALLPAPER. William Morris (English, 1834-1896). Printed by Jeffrey & Co., Islington. England, 1875. Gift of Robert Friedel, 1941-74-20. Photo KP.
3. *"LES JETS D'EAU"* FURNISHING FABRIC. Edouard Bénédictus (French, 1878-1930). Produced by Brunet-Meunie. France, 1925. Jacquard woven cotton and viscose rayon. GAE, 1990-29-2. Photo MTF.
4. FIVE WOODGRAIN WALLPAPERS: Three papers made France, 1800-50. Hand-painted paper. Gift of Josephine Howell, 1972-42-209a/c. One paper probably made USA, c. 1890. Machine-printed paper. Gift of Grace Lincoln Temple, 1938-62-73a. "B100" Wallpaper. Ginbande-Design (Klaus-Achim Heine and Uwe Fischer). Produced by Rasch & Co. Germany, 1992. Machine-printed paper. Gift of Gerrit Rasch and Rasch Tapetenfabrik, 1995-163-7. Photo VS.

Pages 134-135

1. "CALYX" FURNISHING FABRIC. Lucienne Day (English, b. 1917). Produced by Heal Fabrics Limited. England, 1951. Screen-printed viscose rayon. Gift of Eddie Squires, 1992-3-4. Photo DC.
2. FURNISHING FABRIC. Josef Hillerbrand (German, 1892-1981). Produced by the Deutsche Werkstätte, Berlin. Germany, 1926. Block-printed linen. Gift of Teresa Kilham, 1958-88-12. Photo DC.
3. *"THARRKARRE"* FABRIC HANGING. Judith Kngawarreye (Australian). Australia, 1989. Silk batik. GAE and Pauline Cooper Noyes Fund, 1992-21-1. Photo MTF.
4. *"LA TERRE"* FURNISHING FABRIC. Mlle Clarinval. Produced by Tassinari and Chatel, Lyon. France, 1925. Woven silk. Gift of Susan Dwight Bliss, 1931-1-7. Photo MTF.
5. "PERI" FURNISHING FABRIC. Owen Jones (English, 1809-1874). Warner, Sillett & Ramm. England, 1870-71. Woven silk. Sarah Cooper Hewitt Fund, 1995-80-1. Photo DC.

Pages 136-137

1. ARMCHAIR. Probably England, c. 1880. Wood, ivory. Gift of Albert Hadley, 1982-29-1. Photo JW.
2. SIDE CHAIR. Italy, 16th century. Wood. Gift of Charles W. Gould, 1926-11-11. Photo DC.
3. "SACCO" CHAIR. Piero Gatti (Italian, b. 1940), Cesare Paolini (Italian, b. 1937), and Franco Teodoro (Italian, b. 1939). Mfr'd by Zanotta. Italy, 1969. Leather, polystyrene pellets. Gift of ICF, Inc., 1981-59-1. Photo DC.
4. CHILD'S CHAIR. Charles Eames (American, 1907-1978) and Ray Eames (American, 1912-1988). Mfr'd by Herman Miller Furniture Company, Zeeland, MI. USA, 1944. Bent, laminated wood. Gift of Mrs. Eric Larrabee, 1991-144-1. Photo ST.
5. ARMCHAIR. China (for export), c. 1815. Bamboo, cane. Gift of Mrs. William Pedlar, 1962-75-1. Photo DC.
6. ARMCHAIR. Warren McArthur (American, 1885-1961). Mfr'd by Warren McArthur Corp. USA, c. 1935. Aluminum, vinyl upholstery. Gift of Mel Byars, 1991-59-10. Photo DC.
7. ARMCHAIR FOR MUSEUM OF MODERN ART'S MEMBERS' AND TRUSTEES' ROOM. Russel Wright (American, 1904-1976). USA, 1934. Wood, animal hide, upholstery. Gift of Russel Wright, 1976-15-8. Photo MR.
8. ARMCHAIR. Possibly France, c. 1810. Mahogany, gilding, silk upholstery (replaced). Gift of Eleanor and Sarah Hewitt, 1920-15-402. Photo DC.
9. SIDE CHAIR. Attributed to Carlo Zen (Italian, 1851-1918). Italy, c. 1900. Fruitwood with mother-of-pearl, metal, and other inlays; silk upholstery. Gift of Donald Vlack, 1971-49-1. Photo DC.
10. "GRASSHOPPER" ARMCHAIR AND OTTOMAN. Eero Saarinen (Finnish, 1910-1961, active USA). Mfr'd by Knoll. USA, 1945-50. Bent, laminated wood, wool upholstery (replaced). Gift of Mel Byars, 1991-59-59,60. Photo JW.
11. SIDE CHAIRS. From *Le Garde-meuble,* Vol. 79. Paris: D. Guilmard, 1850s. Photo DC.
12. ARMCHAIR. Marcel Breuer (Hungarian, 1902-1981, active USA from 1937). Germany, 1925. Tubular steel, canvas. Gift of Gary Laredo, 1956-10-1. Photo DC.
13. PAIR OF SIDE CHAIRS. René Drouet. Aubusson upholstery design after Madeleine Luka (b. 1900). France, c. 1939. Wood, wool tapestry upholstery. Bequest of Faïe J. Joyce, 1989-23-1,2. Photo JW.
14. SIDE CHAIR FOR THE IMPERIAL HOTEL, TOKYO. Frank Lloyd Wright (American, 1867-1959). USA and Japan, c. 1920. Oak, vinyl upholstery. Gift of Tetsuzo Inumaru, 1968-137-1c.
15. SIDE CHAIR FOR THE PURKERSDORF SANATORIUM, VIENNA. Josef Hoffmann (Austrian, 1870-1956). Mfr'd by Jacob & Josef Kohn. Austria, 1904-06. Wood, leather upholstery. Purchase, combined funds and gift of Crane and Company, 1968-6-1. Photo DK.
16. ARMCHAIR. Warren Platner (American, b. 1919). Mfr'd by Knoll. USA, 1965. Bronzed steel wire, nylon upholstery. Gift of Knoll Associates, Inc., 1971-16-1. Photo DC.
17. SIDE CHAIR. Emile Gallé (French, 1846-1904). France, c. 1900. Various woods, upholstery (replaced). Gift of Mrs. Jefferson Patterson, 1979-54-1. Photo DC.

18. ROCKING CHAIR. Charles Eames (American, 1907-1978) and Ray Eames (American, 1912-1988). Mfr'd by Herman Miller Furniture Company, Zeeland, MI. USA, c. 1950. Fiberglass, metal, wood, rubber. Gift of Barry Friedman and Patricia Pastor, 1986-99-46. Photo JW.
19. SIDE CHAIR. Peter Behrens (German, 1868-1940). Produced for Wertheim Department Store, Berlin. Germany, 1902. Oak, rattan seat (replaced). GAE, 1985-121-1. Photo DC.
20. "EASY EDGES" CHAIR. Frank O. Gehry (American, b. 1929). Mfr'd by Jack Brogan. USA, 1971-72. Corrugated cardboard. Gift of William Woolfenden, 1988-79-2. Photo JW.
21. CORNER LOVESEAT. From *Le Garde-meuble*, Vol. 79. Paris: D. Guilmard, 1850s. Photo KP.

Design for Communicating

Page 138

POP-UP FROM *ALICE'S ADVENTURES IN WONDERLAND.* Lewis Carroll (English, 1832-1898), edited by Ron Stover. NY: Modern Promotions/Publishers, c. 1983. Photo VS.

Pages 140-141

1. EMBROIDERED SAMPLER. Catharine Congdon, aged 10 (American). Newport, RI. USA, 1773. Embroidered silk on linen. Bequest of Gertrude M. Oppenheimer, 1981-28-151. Photo DC.
2. SPECIMENS OF PRINTING TYPES AND ORNAMENTS CAST BY JAMES CONNER & SONS. NY: James Conner & Sons, 1855. Photo MTF.
3. *INTERNATIONAL WOMEN WORKERS DAY* POSTER. Valentina N. Kulagina (Russian, 1902 to mid-1930s). Soviet Union, 1930. Color lithograph on paper. Purchase with gift from Mr. George A. Hearn, 1992-164-1. Photo MTF.

Pages 142-143

1. DESIGNS FOR THREE MONOGRAMS: B.E., E.L., N.S. Jean E. Puiforcat (French, 1897-1945). France, c. 1930-45. Pen, brush, ink, gouache, graphite on illustration board. Gift of John Finguerra in honor of Jim Finguerra, 1995-164-87, 99, 85. Photo VS.
2. DOUBLE-PAGE SPREAD FROM *CATALOG DESIGN.* Knud Lönberg-Holm and Ladislav Sutnar (Czech, 1897-1976, active USA from 1939). NY: Sweet's Catalogue Service, 1944. Photo MTF.
3. DYER'S RECORD BOOK. Compiled by Thomas Ratcliffe. England (near Manchester), 1812-22. Leather, paper, block- and roller-printed cotton. Au Panier Fleuri Fund, 1987-46-1.

Pages 144-145

1. *ENCYCLOPÉDIE, OU, DICTIONNAIRE RAISONNÉ DES SCIENCES, DES ARTS ET DES MÉTIERS.* Plate IX, "*Tapissier*," Vol. IX of the Plates. Denis Diderot (French, 1713-1784). Paris: Briasson, 1751-65. Photo MTF.
2. PLATE. Sergei Chekhonin (Russian, 1878-1936). Mfr'd at the State Porcelain Factory, Petrograd. Soviet Union, 1919. Porcelain, enameled and gilded. The Henry and Ludmilla Shapiro Collection, partial gift and purchase, DAAAF and SRCAPF, 1989-41-10. Photo HI.
3. FRENCH REVOLUTIONARY WALLPAPER. Probably produced by Jacquemart and Bénard, Paris. France, c. 1792. Block-printed paper. Gift of John Jay Ide Collection, 1986-106-1. Photo KP.
4. "EDWARD VIII" COMMEMORATIVE HANDKERCHIEF. England, 1936. Printed cotton. Gift of Edith Wetmore, 1937-27-1. Photo MTF.

Pages 146-147

1. UNMOUNTED FAN LEAF. Antoine Denis Chaudet (French, 1763-1810), Charles Percier (French, 1764-1838), and Pierre Fontaine (French, 1762-1853). Engraved by Jean Godefroy (French, 1771-1839). France, 1797. Printed silk. Bequest of Richard C. Greenleaf, 1962-58-4. Photo DC.
2. *BRISÉ* (FOLDING) FAN FROM VIENNA WORLD EXPOSITION. Austria, 1873. Wood, printed paper. Gift of Mrs. James O. Green, 1920-10-2. Photo MTF.
3. "FRANKLIN D. ROOSEVELT" CIGAR LABEL. USA, c. 1930-35. Letterpress on paper. Gift of Ellery Karl, 1975-74-1(9). Photo MTF.
4. THREE MATCHSAFES. Various manufacturers. USA and England, late 19th century. Silver, enamel, brass. Gift of Carol B. Brener and Stephen W. Brener, 1978-146-28, 1982-23-217, 1978-146-241.
5. *DADA* Poster. Paul Rand (American, b. 1914). USA, 1951. Screen-printed on paper. Gift of Paul Rand, 1981-29-206. Photo MTF.

Pages 148-149

1. "THE GRAND BOOBY." From *The Natural History of Carolina, Florida and the Bahama Islands...*, Vol. I, p. 86. Mark Catesby (English, 1683-1749). London: printed for Benjamin White, 1771. Photo MTF.
2. "*LE SIFILET.*" From *Histoire naturelle des oiseaux de paradis...*, Vol. I, Plate 12. François Le Vaillant (French, 1753-1824). Paris: Chez Denn le jeune... (et) chez Perlet..., 1806. Gift of Robert L. Chanler. Photo MTF.
3. PLATE FROM *STUDIES IN DESIGN.* Christopher Dresser (British, 1834-1904). England: Cassell, Petter & Galpin, 1876, Plate VIII.

Pages 150-151

1. THREE RARE BOOKS. *Wine, Women, and Song...* by John Addington Symonds. London: Chatto and Windus, 1884. *The Durbar* by Mortimer and Dorothy Menpes. London: Adam and Charles Black, 1903. *Andächtiges Geist- und Trostreiches Gebett-Büchlein* by Martin von Cochem (1634?-1712). Cologne: Herman Demen, 1675. Photo VS.
2. *THE BOOKPLATES & MARKS OF ROCKWELL KENT.* NY: Pynson Printers for Random House, 1929, images 48-49. Gift of William Howle, 1990. Photo DC.

Pages 152-153

1. BISCUIT TIN IN THE FORM OF A STACK OF BOOKS. Produced by Huntley & Palmer's, Ltd. England, 1901. Tin. Gift of Alvin and Eileen Preiss, 1990-93-3. Photo JW.
2. PHOTOGRAPH OF BRODARD & TAUPIN STOREFRONT. Paris, c. 1924. Thérèse Bonney Collection, ASF117B. Gift of Thérèse Bonney, 1939.

Pages 154-155

1. WOMAN AT A WRITING DESK. From *L'Art d'Ecrire*, Plate 3. France, 18th century. Engraving. Photo MTF.
2. *POST OFFICE* POSTER. For the British General Post Office. Edward McKnight Kauffer (American, 1890-1954, active England from 1914). England, November 5, 1936. Gouache, graphite on paper. Gift of Mrs. E. McKnight Kauffer, 1963-39-523.
3. PELIKAN-FÜLLHALTER POSTER. William Metzig (German, 1893-1989, active USA after 1939). Hanover, Germany, 1930. William Metzig Collection. Photo MTF.

Pages 156-157

1. PLATE FROM *ALBUM DES ARTS UTILES ET AMUSANTS.* Compiled by various artists and amateurs. Paris: Dentu, 1832. Lithograph, hand-colored with watercolor. Purchase in memory of Erskine Hewitt, 1952-144-1. Photo MTF.
2. ALPHABET BOWL. Eric Ravilious (English, 1903-1942). Mfr'd by Wedgwood, Staffordshire. England, 1937. Earthenware with transfer-printed decoration. Gift of Paul F. Walter, 1992-42-2. Photo HI.
3. *DADA* POSTER. Paul Rand (American, b. 1914). USA, 1951. Screen-printed on paper. Gift of Paul Rand, 1981-29-206. Photo MTF.
4. DOUBLE-PAGE SPREAD FROM *CATALOG DESIGN.* Knud Lönberg-Holm and Ladislav Sutnar (Czech, 1897-1976, active USA from 1939). NY: Sweet's Catalogue Service, 1944. Photo MTF.

Pages 158-159

1. PLATE FROM *ALPHABETE=ALPHABETS =ALPHABETS= LETTERLIJSTEN= ALFABETI=ALFABETOS.* Attributed to Friedrich Christmann and Hermann Junker (1838-1899). Frankfurt-am-Main: B. Dondorf, 1870s, Plate 39. Photo MTF.
2. "MORO LIGHT: FABRICA DE TABACOS" CIGAR LABEL. USA, c. 1916. Chromolithograph on paper. Gift of Ellery Karl, 1975-74-1(16). Photo MTF.
3. "B" PRINTING PLATE. Europe or USA, c. 1870. Brass. Gift of Mrs. William A. Hutcheson, 1943-31-15. Photo HI.
4. WALLPAPER SAMPLES. From a factory workbook. Jacquemart and Bénard, Paris. France, 1824. Block-printed on paper. DAAAF, 1986-34-2. Photo SH.

Pages 160-161

1. *ENCYCLOPÉDIE, OU, DICTIONNAIRE RAISONNÉ DES SCIENCES, DES ARTS ET DES MÉTIERS.* Plate IX, "*Tapissier*," Vol. IX of the Plates. Denis Diderot (French, 1713-1784). Paris: Briasson, 1751-65. Photo MTF.
2. PATTERN SAMPLE PLATE. Probably France, c. 1875. Enameled, gilded porcelain. DAAAF, 1990-152-1. Photo JW.
3. *THE PRACTICAL OSTRICH FEATHER DYER.* Alexander Paul; revised and corrected by Dr. Morris Frank. Philadelphia: Published by Mrs. Morris Frank, 1888, p. 46a. Photo MTF.
4. TEXTILE DESIGN. Louis-Albert DuBois (Swiss, 1752-1818). For Fabrique de Fazy aux Bergues, Geneva. Switzerland, c. 1801. Pen and ink, gouache, watercolor over graphite on paper. Friends of the Museum Fund, 1957-46-19.
5. SALESMAN'S SAMPLE CARD OF PRINTED FABRICS. France, early 19th century. Cotton, paper. Alice Baldwin Beer Memorial Fund, 1986-19-1.
6. PLATE FROM *STUDIES IN DESIGN.* Christopher Dresser (British, 1834-1904). England: Cassell, Petter & Galpin, 1876, Plate VIII.
7. WEAVER'S RECORD BOOK. Compiled by the Jackson Family. England, in continuous use from the late 17th to late 18th centuries. Leather, paper, ink. Purchase in memory of Mrs. Clarence Webster, 1958-30-1. Photo MTF.
8. UNFINISHED EMBROIDERY. England, 17th century. Silk, silk-wrapped wire, and mica on silk. Gift of Sarah Cooper Hewitt, 1903-11-28. Photo DC.

Pages 162-163

1. *BRISÉ* (FOLDING) FAN FROM VIENNA WORLD EXPOSITION. Austria, 1873. Wood, printed paper. Gift of Mrs. James O. Green, 1920-10-2. Photo MTF.
2. "*LA BALON DE GONESSE*" FURNISHING FABRIC. Produced by Manufacture Royale de Oberkampf, Jouy. France, 1784. Printed cotton. Au Panier Fleuri Fund, 1961-116-5. Photo DC.

3. COSMONAUTS AND ROCKET. Mfr'd by Gzhel Porcelain Factory. Soviet Union, 1960-70. Porcelain with enamel decoration. The Henry and Ludmilla Shapiro Collection, Gift of Ludmilla Shapiro, 1993-13-1. Photo HI.
4. "LUNAR ROCKET" FURNISHING FABRIC. Eddie Squires (English, 1940-1994). Produced by Warner Fabrics, London. England, 1969. Screen-printed cotton. Gift of Eddie Squires, 1991-102-1. Photo DC.
5. "BOAT RACE DAY" VASE. Eric Ravilious (English, 1903-1942). Mfr'd by Wedgwood, Staffordshire. England, 1938. Earthenware with printed decoration. Gift of Paul F. Walter, 1989-110-1. Photo HI.

Pages 164-165

1. "THE REDS AND THE WHITES" CHESS SET. Natalya Danko (Russian, 1892-1942). Mfr'd by Lomonosov Porcelain Factory. Soviet Union, introduced 1922. Gift of the estate of Harrison E. Salisbury, 1994-122-5. Photo VS.
2. SET OF BUTTONS. Painted decoration probably by Agostino Brunias (Italian, active England and the West Indies, 1730s-1810s?). Made in England or Haiti, c. 1796. Painted canvas, ivory, glass, metal. Gift of R. Keith Kane from the estate of Mrs. Robert B. Noyes, 1949-94-1/18. Photo VS.
3. VASE WITH PORTRAIT OF STALIN. Frontispiece from *Catalogue of Porcelains, Faiences, and Majolica.* Ukrainian National Commissariat. Kiev: Ukrainian State Publishing House, 1940. Photo DC.

Pages 166-167

1. FRENCH REVOLUTIONARY WALLPAPER. Probably produced by Jacquemart and Bénard, Paris. France, c. 1792. Block-printed paper. Gift of John Jay Ide Collection, 1986-106-1. Photo KP.
2. PLATE. Sergei Chekhonin (Russian, 1878-1936). Mfr'd at the State Porcelain Factory, Petrograd. Soviet Union, 1919. Porcelain, enameled and gilded. The Henry and Ludmilla Shapiro Collection, partial gift and purchase, DAAAF and SRCAPF, 1989-41-10. Photo HI.
3. UNMOUNTED FAN LEAF. Antoine Denis Chaudet (French, 1763-1810), Charles Percier (French, 1764-1838), and Pierre Fontaine (French, 1762-1853). Engraved by Jean Godefroy (French, 1771-1839). France, 1797. Printed silk. Bequest of Richard C. Greenleaf, 1962-58-4. Photo DC.
4. "EDWARD VIII" COMMEMORATIVE HANDKERCHIEF. England, 1936. Printed cotton. Gift of Edith Wetmore, 1937-27-1. Photo MTF.
5. PITCHER. Probably mfr'd by H. Baggeley. England, 1855. Enameled and gilded porcelain. DAAAF, 1989-63-1. Photo DC.

Pages 168-169

1. *POWER—THE NERVE CENTRE OF LONDON'S UNDERGROUND* POSTER. Edward McKnight Kauffer (American, 1890-1954, active England from 1914). England, 1930. Color lithograph. Gift of Mrs. E. McKnight Kauffer, 1963-39-45.
2. *RURAL INDUSTRIES* POSTER. For Rural Electrification Administration, U.S. Department of Agriculture. Lester Beall (American, 1909-1969). USA, c. 1937-41. Offset lithograph and screenprint on paper. SRCAPF, GAE, Friends of Drawings and Prints, Sarah Cooper Hewitt Fund, and through the gift of Mrs. Edward C. Post, 1995-106-1. Photo KP.
3. *USA.* Cover for catalogue for XXXII Biennial International Exposition of Art, Venice. Elaine Lustig Cohen (American, b. 1927). USA, 1964. Offset lithograph on paper. Gift of Tamar Cohen and David Slatoff, 1993-31-62. Photo DC.
4. *EL DOMINIO DEL FUEGO* POSTER. Luis Vega (Cuban, b. 1944, active USA). Cuba, 1972. Screenprint on paper. SRCAPF and through the gift of anonymous donors, 1994-65-13. Photo MTF.

Pages 170-171

1. "GRAPHIC FORM" DOUBLE-PAGE SPREAD. From *International Dictionary of Symbols.* Henry Dreyfuss (American, 1904-1972). Dummy book, December 12, 1969; revised and published as *Symbol Sourcebook: An Authoritative Guide to International Graphic Symbols,* 1972. Henry Dreyfuss Collection. Gift of Doris and Henry Dreyfuss, 1972. Photo MTF.
2. PRINTER'S PROOF FOR LUMIUM LIMITED LETTERHEAD. Edward McKnight Kauffer (American, 1890-1954, active England from 1914). England, 1935. Printer's proof. Gift of Mrs. E. McKnight Kauffer, 1963-39-690c.
3. *FASHIONS IN FLIGHT* POSTER. For American Airlines. Edward McKnight Kauffer (American, 1890-1954, active England from 1914). USA, c. 1947. Color lithograph on paper. Gift of Mrs. E. McKnight Kauffer, 1963-39-132. Photo MTF.
4. MONOWATT LOGO. William Metzig (German, 1893-1989, active USA after 1939). USA, probably 1940s. William Metzig Collection. Photo DC.
5. DESIGNS FOR SIX MONOGRAMS: E.L., N.S., I.J., E.D., B.E., O.T. Jean E. Puiforcat (French, 1897-1945). France, c. 1930-45. Pen, brush, ink, gouache, graphite on illustration board. Gift of John Finguerra in honor of Jim Finguerra, 1995-164-99, 85, 98, 86, 87, 109. Photo VS.

Pages 172-173

1. PHOTOGRAPH OF CHEER DETERGENT BOXES. For Proctor and Gamble. Donald Deskey Associates. USA, c. 1957. Donald Deskey Collection. Gift of Donald Deskey Associates.
2. SIX MATCHSAFES. Various manufacturers. USA, c. 1900. Metal, celluloid, paper. Gift of Carol B. Brener and Stephen W. Brener, 1978-146-68, 1978-146-21, 1980-14-1423, 1980-14-1420, 1980-14-1422, 1978-146-257. Photo VS.
3. BANDBOX. Mfr'd by Putnam & Roff, Paper Hanging and Band Box Manufacture, Hartford, CT. USA, 1821-24. Block-printed paper on cardboard. Gift of Mrs. Frederick Thompson, 1913-45-10. Photo DC.
4. FIORUCCI SHOPPING BAG. Fiorucci Graphics Studio. Italy, c. 1977. Offset color lithograph on paper. Photo MTF.
5. PROMOTIONAL FABRIC FOR BASSETT'S CANDIES. England, late 1930s-early 1940s. Roller-printed rayon. Gift of Eddie Squires, 1992-3-2. Photo DC.

Pages 174-175

1. 1939 NEW YORK WORLD'S FAIR PEEP SHOW. USA, 1939. Paper. Photo VS.
2. *CALIFORNIA INSTITUTE OF THE ARTS* POSTER. Jayme Odgers (American, b. 1939) and April Greiman (American, b. 1948). USA, 1978. Offset lithograph on paper. Gift of April Greiman, 1981-29-25. Photo MTF.
3. *DEATH OF A SALESMAN* POSTER. Fahrenheit (Paul Montie, American, b. 1965, and Carolyn Montie, American, b. 1966). USA, 1993. Offset lithograph on paper. Gift of Paul Montie and Carolyn Montie, 1996-3-1. Photo KP.
4. *THE EDGE OF THE MILLENNIUM* POSTER. For a National Design Museum symposium. Lorraine Wild (b. 1953) with Reverb. USA, 1991. Offset lithograph on paper. Photo DC.

Pages 176-177

1. COVER OF *MADEMOISELLE,* MARCH 1950. From *The Art of Graphic Design* by Bradbury Thompson (American, 1911-1995). New Haven, CT: Yale University Press, 1988, p. 173, fig. 229. Photo MTF.
2. PROPOSED COVER FOR *THE NEW YORKER* or *PROMENADE MAGAZINE.* Christina Malman (American b. England, 1912-1959). USA, c. 1939. Graphite, pen and ink, watercolor, crayon on paper. Gift of Christina Malman, 1947-110-3.

Pages 178-179

1. VELVET PANEL. Iran, early 17th century. Woven velvet with silk and metallics. Anonymous gift, 1977-119-1.
2. "WEDDING" WALLPAPER. Saul Steinberg (American, b. 1914). Produced by Piazza Prints, Inc., NYC. USA, 1950. Screen-printed paper. Gift of Harvey Smith, 1950-126-1. Photo KP.

Pages 180-181

1. COVER OF *THREE HUNDRED AESOP'S FABLES.* London: George Rutledge and Sons, 1867. Photo MTF.
2. DALMATIAN ENAMELED BOX. England, c. 1850. Enameled, gilded copper. Gift of Sarah Cooper Hewitt, 1931-84-23.
3. TOY GRASSHOPPER. Europe, c. 1900. Metal, acetate. Anonymous gift, 1949-49-9. Photo DC.
4. "TOUCAN WITH YELLOW COLLAR." From *Histoire naturelle des oiseaux de paradis...,* Vol. I, Plate 2. François Le Vaillant (French, 1753-1824). Paris: Chez Denn le jeune... (et) chez Perlet..., 1806. Gift of Robert L. Chanler. Photo MTF.

Pages 182-183

DOUBLE-PAGE SPREAD FROM *LOCUPLETISSIMI RERUM NATURALIUM THESAURI....* Albert Seba (Dutch, 1665-1736). Amsterdam: Janssonio-Waesbergios [etc.], J. Wetstenium & Gul. Smith, 1734-65. Photo VS.

Front endpaper

"BUTTONS" WALLPAPER. Mfr'd by Piazza Prints, Inc., NYC. USA, 1952. Screen-printed. Gift of *The Wallpaper Magazine,* 1953-51-3.

Back endpaper

"WOODPIDGEON" WALLPAPER. Edward Bawden (English, 1903-1989). Mfr'd by The Curwen Press. England, 1927. Lithograph after a linocut. SRCAPF and Pauline C. Noyes Fund, 1996-32-3. Photo AG.

The Curatorial Departments

Wallcoverings

The Museum holds the largest and most comprehensive survey of wallcoverings in the country with over 10,000 examples. Collected for their historic associations, for design inspiration, or as examples of printed material, the holdings are especially rich in examples of exuberant French floral compositions from the 19th century, William Morris-inspired patterns, and early American block-printed papers. Wallcoverings made for the most fashionable homes and the simplest of cottages are represented, as are panoramic and scenic papers, 17th-century Dutch gilded leather wallcoverings, and 18th-century French block-printed arabesques. Twentieth-century wallcoverings are represented in papers designed by students at the Bauhaus in 1929 and by Frank Lloyd Wright in 1956, and in such recent additions as Jhane Barnes's 1992 designs made with acrylic resin powders. The holdings also include sample books, printing blocks, and advertisements.

Applied Arts and Industrial Design

The Department of Applied Arts and Industrial Design is home to more than 30,000 three-dimensional objects that date from antiquity to the present. Chairs are a special area of strength in the furniture collection; the collection of lighting fixtures is also one of rare depth. The Decloux Collection of carved wooden wall panels, dating from the Renaissance to the 19th century, is an outstanding resource for historic interior styles and ornament. Silver tablewares, wrought iron gates and architectural elements, and a large selection of French 18th-century gilt-bronze furniture mounts can be found in the metalwork collection. The Museum's comprehensive ceramics collection ranges from ancient Greek vessels and pre-Columbian pottery to 20th-century tablewares. Notable in the glass collection are Syrian and Roman pieces, and Irish cut-glass tablewares from the early 19th century. Other highlights of the Applied Arts Department are the Metzenberg Collection of historic cutlery, the Shapiro Collection of 20th-century Soviet porcelains, the Brener Collection of match-safes, Japanese *tsuba* (sword fittings), jewelry, buttons and fasteners, precious small boxes and cases, birdcages, lacquer and leather work, enamels, plastics, and models and prototypes for cars and other products. Appliances represented in the department's collections include machines and tools made for home and office.

Textiles

An extraordinarily wide range of textile techniques is represented in the 30,000 pieces of the Museum's textile collection—embroidery, knitting, crochet, braiding, knotting, quilting, and needle- and bobbin-made lace, as well as printing (using engraved plates, wooden blocks, rollers, lithography, and silk screens), various forms of resist dyeing (including tie-dye, ikat, batik, and painted, printed, and stenciled resist), fabric painting, and weaving (ranging from simple plain weave to complex drawloom and jacquard woven patterns). The date range of the collection is just as far-reaching. The earliest pieces are from Han Dynasty China (206 B.C.-A.D. 221) and pre-Columbian South America, while the latest pieces were made in the 1990s. Special strengths of the department are woven European silks from the 13th through the 18th centuries, French and English printed fabric from the 18th and early 19th centuries, the collection of nearly 1,000 embroidered samplers, and classic European laces from the 16th and 17th centuries.

Drawings and Prints

Housing over 160,000 works on paper, the National Design Museum's Drue Heinz Study Center for Drawings and Prints ranks among the world's foremost repositories of European and American designs for architecture, decorative arts, gardens, interiors, ornament, theater, textiles, and graphic and industrial design. Its encyclopedic holdings range from a 14th-century North Italian textile design and a late 15th-century German drawing for a Gothic steeple to the work of such contemporary American and European designers as Donald Deskey, Henry Dreyfuss, Jean Puiforcat, Edward McKnight Kauffer, Dakota Jackson, Eva Zeisel, Elaine Lustig Cohen, and Robert Wilson. No other collection in this country matches its strength in 17th- to 19th-century Italian and French drawings and prints pertaining to ornament, decorative arts, and architecture. Other treasures include the rare group of early 19th-century drawings by Frederick Crace for the exotic interiors of the Royal Pavilion at Brighton, and over 400 Japanese *kata-gami* (stencil) patterns for textiles. Twentieth-century American designs for architecture and industrial design are also represented in considerable depth, along with 20th-century graphic design, comprising a variety of print media, from posters to book covers to stationery.

Library and Archive

The Doris and Henry Dreyfuss Memorial Study Center, a branch of Smithsonian Institution Libraries, contains 55,000 volumes, including 5,000 rare books pertaining to ornament, architecture, and decorative arts. The library's collections focus on interior, graphic, and industrial design, along with books on textiles, wallcoverings, architecture, and design. The Library has significant holdings of swatch books, trade catalogues, world's fair literature, children's books, pattern books, and over 600 pop-up and moveable books. The National Design Museum's Archives contain the papers, promotional materials, clippings, and photographs of designers and design firms including M&Co., Edward F. Caldwell, George Nathan Horwitt, Don Wallance, Donald Deskey, Henry Dreyfuss, and Ladislav Sutnar. The Archives also contain special resource files on African-American and Latino-Hispanic designers. The Library and Archives are open to the public by appointment.